CALIFORNIA'S
CHANGING
LANDSCAPES

PREFACE

WE STARTED TO WRITE THIS BOOK IN 1982, TEN YEARS BEFORE publication. In that time, the population of California rose fifty percent, from twenty million to thirty million. The multiple demands on our state's land increased just as steeply: demands for living space, farm and pasture space, recreational space, and wood products such as lumber and fuel. California's population is expected to double—to sixty million—in the twenty years between 1992 and 2012. It is time to step away from this pace, to take a measure of what we have, to compare it with what we once had, and to decide what we want for the future.

This book is a guide for that step-away process. We write about the vegetation that clothes the landscapes of California: its diversity and conservation. The intimate relationships between people, vegetation, and the landscape are of vital importance to us because we are embedded in that vegetation. Although we may not be conscious of it daily, the quality of our lives is as intimately affected by vegetation now as it was more than 12,000 years ago when Native Americans first came to this unique and diverse slice of the earth. This book describes the landscapes and plant resources which were once here, how they came to be consumed, what has replaced them, and what might remain in the future. The quality of that future will be enhanced if the remaining natural vegetation of California is protected and its disjointed portions are restored. We hope that reading this book will stimulate you to help realize that kind of future.

Although we fill this book with descriptions of nature in California, the political boundaries of California are not themselves natural. The area we call California is special because so many biotic domains come jumbled together here within the artificial boundaries of the state: the wet forests of the Pacific Northwest; the cold, dry sagebrush-grassland northern plains; the hot deserts of Mexico; and the summer dry-winter wet coastal, valley, and foothill vegetation of the Californian biotic province. As poet and author Gary Snyder has written, California is "after all, a recent human invention—with hasty straight-line boundaries that were drawn with a ruler on a map and rushed off to an office in D.C." Nevertheless, he adds, "I am not arguing that we should redraw the boundaries of the jurisdiction called California, although that could happen some far day. But we are looking at certain long-range realities. I do think it leads toward the next step in the evolution of human citizenship on the North American continent. . . People are coming to the realization that we can become members of ancient biological communities in a different kind of citizenship."

Many people contributed to this book. The readability and flow of the text were improved enormously by the careful editing of Valerie Whitworth, Robert Ornduff, Phyllis Faber, Barbara Leitner, and several anonymous reviewers. We co-authors accept full responsibility for errors of fact or awkwardly written passages that still may remain. Our ethnobotanical section was immeasurably improved by contributions from Kat Anderson. Outstanding photographs of vegetation were provided by Kat Anderson, Walt Anderson, David Cavagnaro, Dan Cheatham, Susan Cochrane, Sonia Cook, Bob Haller, Verna Johnston, Todd Keeler-Wolf, Oren Pollak, and The Nature Conservancy. Drawings of native California bunchgrasses were provided by Jean Brennan-Hagen; other illustrations were created by Michal Yuval. Final manuscript preparation was accomplished on a Macintosh computer and LaserWriter, courtesy of the Calaveras Big Trees Association.

Plant names in the text are based on Munz and Keck's 1963 *Flora of California,* augmented by our perception of the most widely accepted or appropriate common names. The 1993 *Jepson Manual* includes many changes for both common and scientific names. To accommodate these changes so late in the process of publication we have cross-listed them (as synonyms) in the index of plant names at the end of the book. We thank Barbara Leitner for her enormous editorial service in compiling this list from pre-publication galleys and proofs of the new *Jepson Manual.*

Michael Barbour
Arnold, California, December 1992

CALIFORNIA'S CHANGING LANDSCAPES

Diversity and Conservation of California Vegetation

MICHAEL BARBOUR, BRUCE PAVLIK,
FRANK DRYSDALE,
AND SUSAN LINDSTROM

CALIFORNIA NATIVE PLANT SOCIETY
SACRAMENTO, CALIFORNIA

©1993 Barbour, M., B. Pavlik, F. Drysdale and S. Lindstrom.
Published by the California Native Plant Society. All rights
reserved. No part of this book may be reproduced in any form or
by any electronic or mechanical means without permission of the
publisher. Requests should be made to CNPS, 1722 J Street,
Suite 17, Sacramento, CA 95814.

Editor: **Phyllis Faber**
Contributing Editors: **Barbara Leitner, Nora Harlow**
Designer: **Beth Hansen**
Typesetting / Linotronic: **Beth Hansen / Canterbury Press**
Color Conversions: **Sieg Photographics**
Color Separations: **Colortec**
Printing: **Lorraine Press**

This book was set in Goudy and printed on 80# Sterling Gloss.

Printed and bound in the United States.
ISBN 0-943460-17-4
Library of Congress Catalog Card Number: 92-076201

---

*Cover: Grassland wildflowers in spring, northern Los
Angeles Co.*

---

*Back cover: Soil erosion on Santa Catalina Island caused
by overgrazing from goats, pigs and deer.*

---

*Title page: Annual grasslands of Merced Co., golden in
summer and fall.*

CONTENTS

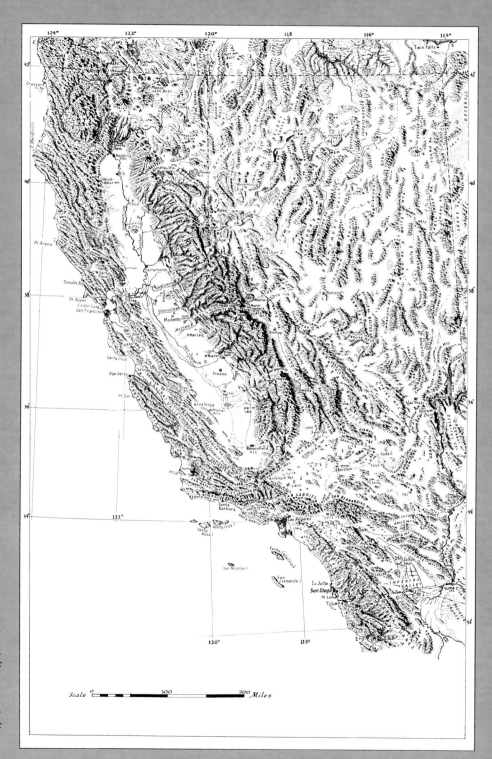

A landform map of
California showing its
major topographic features
of mountains and river
drainages. Map courtesy of
Erwin Raisz.

FOREWARD

KNOWLEDGE OF PLANTS AND THEIR LANDSCAPES IS TOO OFTEN taken to be a territory of specialists. Yet such knowledge has been, over most of human history, a wonderfully intimate and common information, shared within families and villages, and as such often is strikingly detailed and accurate. Common knowledge is a measure of knowing and loving a place. Most of us here in California are relative newcomers, and the thought that we might come to know its plants, climates, and peculiar beauties is a fresh and slightly unsettling idea to many. Acquaintance with the primary ecological systems is a first step in becoming authentic inhabitants of any landscape on earth.

We call this western slope of land at the eastern edge of the ocean California. With five or so zones of life and climate, lapping over into high cold deserts and low hot deserts, into rainforest and into the snowfields, it contains one of the richest gatherings of plant species in the world. This book makes our vast and complex landscape available to us. It does so with elegance and economy. It also does not hesitate to argue the case for conservation and restoration, and to present the problems that beset each region. Its vision of a California whose native plant populations do not become entirely degraded, and whose grasslands and forests are sustainable, is visionary but not impractical. It speaks to the good sense of real people in a real place, who expect to continue living there.

The chapter on Native Americans—Native Californians—and the vegetation of California is invaluable. It gives us a clue to the imaginative and instructive loop that leads from the deeply grounded lore of the native inhabitants to the zoning and management practices we might want for our long future. I hope that *California's Changing Landscapes* will encourage people everywhere to join the local community and get to know their non-human neighbors. Long live those magnificent landscapes!

Gary Snyder

1

CALIFORNIA VEGETATION:
Diversity and Change

CALIFORNIA IS RICH. IT FAR EXCEEDS THE SUM OF OUR MYTHS, memories, and experiences. To the Native Californians it was home, it was the world, it provided everything. To others, coming from different directions and at different times, California offered lush pastures, precious metals, tall timber, a new life, a fresh start. It was and continues to be, all things to all people, endlessly accommodating, an inspired landscape offering prosperity arising from diversity.

The economic wealth of California is an expression of its natural diversity. High natural diversity is expressed in California on several scales: within a single local habitat, among a cluster of adjacent habitats all belonging to the same region, and among neighboring regions that clothe the entire state's landscape. The variety of geological substrates, topography, climatic types, soils, vegetation, and species of plants is exceptional. Elevations extend from the second highest (14,495 feet) to the lowest (-282 feet) in the United States. Climates and vegetation range from foggy, humid, dripping wet, shaded coastal forest to arid, hot, open desert scrub. California has examples of all eleven of the world's major soil groups. A half-dozen major North American biomes squeeze together in close juxtaposition here, bringing with them nearly 1,000 distinctive plant communities—groups of species that live together in unique habitats.

Plants that compose those communities vary from small, floating duckweeds to giant sequoias (*Sequoiadendron giganteum*); from frail herbs which live for only weeks to ancient bristlecone pines (*Pinus longaeva*); from recent weedy arrivals to those with fossil records extending back millions of years. California has more than 5,000 kinds of native ferns, conifers, and flowering plants. Japan, with a similar area, has far fewer species. Even with four times California's area, Alaska does not match California's plant diversity, and neither does all of the central and northeastern United States and adjacent Canada combined. Moreover, about thirty percent of all California's native plants are found nowhere else in the world.

When economic prosperity is manufactured from natural diversity, change in the landscape is inevitable. California has become a place of change. The Califor-

A lush carpet of native poppies and grasses—part of the natural wealth of California.

nia we see today is dramatically different from the one Sir Francis Drake claimed for England in 1579, and from the one Father Junipero Serra claimed for Christianity two centuries later. The early Europeans did not see beauty in wild California because they came from tamed landscapes. From offshore, Drake judged harshly the summertime Big Sur coast with these words:

> How unhandsome and deformed appeared the face of the earth itself! . . . shewing trees without leaves, and the ground without greenness in those months of June and July.

Uncomfortable with this strange wilderness, the Europeans began to rear-range the land into more familiar patterns. Within two centuries, woodlands have been cut and replaced by orchards, highways, and cities; marshes have been drained and converted into farms and airports; redwood (*Sequoia sempervirens*) forests are now Douglas-fir (*Pseudotsuga menziesii*) forests, and Douglas-fir forests are now tanbark (*Lithocarpus densiflorus*) and madrone (*Arbutus menziesii*) thick-ets; montane conifer forests have been converted into chaparral, and chaparral into grassland; California prairie has become farms and cities; sand dunes have become city parks, neighborhoods, and golf courses. All this change was brought about with the energy and optimism and anthropocentric energy that typified nineteenth-century America. Walt Whitman's 1873 Song of the Redwood Tree describes California this way:

During spring, grasslands can be dominated by showy, annual species of herbs, such as California poppy, goldfields, and lupine (near Gorman, Los Angeles Co.).

The flashing and golden pageant of California,
The sudden and gorgeous drama, the sunny and ample lands,
The long and varied stretch from Puget sound to Colorado south,
Lands bathed in sweeter, rarer, healthier air, valleys and mountain
    cliffs,
The fields of Nature long prepared and fallow, the silent, cyclic chemistry,
The slow and steady ages plodding, the unoccupied surface ripening,
the rich ores forming beneath;
At last the New arriving, assuming, taking possession,
A swarming and busy race settling and organizing everywhere,
Ships coming in from the whole round world, and going out to the whole
    world,
To India and China and Australia and the thousand island paradises of
    the Pacific,
Populous cities, the latest inventions, the steamers on the rivers, the
    railroads, with many a thrifty farm, with machinery,
And wool and wheat and the grape, and diggings of yellow gold.

However, converting land to serve human purposes invariably erodes natural diversity. This erosion is expressed by the loss of plant and animal species.

Approximately 675 native plant species have become rare or endangered and another thirty-nine have become extinct in the past 200 years. Their unique combinations of genes, slowly produced in eons of evolutionary process, were extinguished in seconds of geologic time. Taking their place are more than 1,000 alien plant species—many noxious weeds—which have been introduced to California.

The erosion of diversity is also expressed by the loss of natural vegetation—whole collections of plant species that share a habitat. Early in this century pioneer plant ecologist Frederick Clements advanced the notion that vegetation is like an organism—not a collection of independent individuals, but an interdependent community with a life of its own. Today, British biochemist James Lovelock's Gaia hypothesis has extended the organism theory to a global level. He contends that the entire biosphere functions as a unified whole—a super-organism—that is able to maintain the environment necessary to sustain life.

Although these views may be extreme, it is true that the destruction of natural vegetation produces drastic and often irreversible changes in the landscape that affect humans directly: rivers flood, livestock starve, wildlife populations diminish, soils become sterile, forests stop regenerating, and wildfires rage. All this has occurred in California, so today we look upon damaged landscapes whose diversity, resiliency, and sustainable productivity are greatly diminished. Reversing this trend will require an understanding of natural vegetation.

Cross-section through Sierran conifer forest vegetation, with needleleaf trees in the overstory, broadleaf trees in the tall understory, shrubs in the short understory, and herbs scattered over the forest floor.

Understanding Vegetation

Vegetation is the plant cover of a region, the clothing over the land. This thin cloth is at once durable and fragile, able to repair and reproduce itself for centuries if the environment remains stable, but subject to irreversible unraveling when environmental stresses become too severe. When vegetation is disrupted, its integrity is fractured. The degraded cover loses ecological relationships, nutrients, and diversity—both plant and animal.

Vegetation plays an essential role in the global movements of water, carbon, oxygen, sulfur, phosphorus, and nitrogen. When vegetation is removed, the movement of these substances is altered, and human populations feel the result. For example, hydraulic mining for gold in nineteenth-century California was a crude mining practice which directed erosive jets of water onto mountain slopes,

destroying the landscape much like a passing glacier removes vegetation and topsoil. Denuded hillsides eroded quickly, creating abnormal sediment loads and devastating floods far downstream in the Central Valley. Sacramento, the state capital, was flooded nearly every year during the Gold Rush days. The Sacramento River changed its course in many places, leaving backwater oxbows and stranding settlements past which the main current had once run. State legislators outlawed hydraulic mining in 1875. Although sediment loads in the Sacramento-San Joaquin River Delta had returned to normal levels by 1950, recovery of the vegetation is incomplete to this day. Revegetation of mine spoils is slow and has resulted in stunted scrubby forests.

As another example, the King, Kaweah, White, and Tule rivers in the San

Ground-dwelling animals, such as black-tailed deer(above), utilize the understory layers of conifer forest vegetation for food and shelter.

Looking up (left) through layers of the Sierran conifer forest, dominated by firs and pines in the overstory.

Joaquin Valley were dammed and their waters were diverted to agricultural purposes in the early twentieth century. Water no longer reached the rich 700-square-mile Tulare Lake near the present site of Bakersfield. The lake dried up because of the loss of incoming water. The 225 square miles of wetlands surrounding the lake also have vanished, removing the lush populations of turtle, fish, beaver, otter, and elk, now gone forever.

The Architecture of Vegetation

We have likened vegetation to a covering over the body of the earth, but the covering is not a simple, thin, homogeneous fabric: it has several layers. The plants which grow together have different sizes, life spans, root systems, and leaf traits. When one sees a forest, all of these layers and characters are mixed and fixed in memory to give a mental picture of forest.

A conifer forest in middle elevations of California mountains has four layers of plants. Small herbs, visible only during the growing season, are scattered over the ground. Common herbs include violets, orchids, calochortus lilies, and bunchgrasses. Each fall the herbs die back to underground stems and roots, so they are not visible until the following spring. In the same layer are prostrate woody plants, such as pinemat manzanita (*Arctostaphylos nevadensis*) or squaw carpet (*Ceanothus prostratus*). The woody plants retain their leaves and stems all through winter, even when covered by several feet of snow.

Tall shrubs make up a second layer of plants: greenleaf manzanita (*Arctostaphylos patula*), bear clover (*Chamaebatia foliosa*), and dwarf tanbark oak (*Lithocarpus densiflorus* var. *echinoides*). The shrubs are typically evergreen, retaining their leaves through winter.

A third layer is composed of understory broadleaf trees. Mountain dogwood (*Cornus nuttallii*), California black oak (*Quercus kelloggii*), and hazelnut (*Corylus cornuta* var. *californica*) are dramatically visible in spring when they flower or in fall when their leaves turn gold, red, and purple.

Finally, the overstory conifer layer, with evergreen, needle-like leaves, dominates all. Conifer trees are said to dominate the vegetation because they are the most abundant growth form in the tallest layer. This layer is not completely closed—that is, the leafy canopies of adjacent trees do not all meet and interlock. There are patches of open space between some trees. Looking down on the canopy from above, only sixty to seventy percent of the ground is seen to be covered by the overstory conifers. Plants of the understory tree, shrub, and herb canopies take advantage of light that passes through gaps between the trees, so when all four layers are looked at from above, 100 percent of the ground is covered.

The architecture of vegetation is defined by the size of plants which make it up, the number of canopies stacked vertically, the growth forms of the plants (tree, shrub, herb), and the leaf traits (evergreen or deciduous; needle-like or broadleaf). Vegetation like the forest just described is found in many parts of California. It is called a mid-montane conifer forest. In California, this term refers to a open forest dominated by conifers with scattered understory trees, shrubs, and herbs, all growing at middle elevations of mountain ranges high enough to receive winter snow but low enough to have a long, warm, dry growing season.

Mid-montane conifer forest is the name of one vegetation type; California has dozens of others. Each vegetation type's name includes a geographic term and the dominant plant growth form. From place to place, the identity of the species making up the vegetation may change, but the structure of the vegetation is constant. Names of other California vegetation types include upper montane conifer forest, central valley annual grassland, alpine tundra, coastal salt marsh, warm desert scrub, coastal temperate rainforest, and northern coastal scrub.

A vegetation type may contain many communities, each community differing in the species that make up the dominant plant growth form. In contrast to vegetation types, communities are usually named after their dominant species. Examples of communities within the mid-montane conifer forest vegetation include ponderosa pine (*Pinus ponderosa*) forest, Douglas-fir forest, white fir (*Abies concolor*) forest, mixed conifer forest, and giant sequoia forest.

A major difficulty in summarizing California's vegetation cover is developing a consistent classification system. Botanists have used different schemes to classify the region's plant cover. Some use names that emphasize the dominant plants, for example redwood forest, or pinyon (*Pinus monophylla, P. edulis, P. quadrifolia*)-juniper woodland. Some systems use physical features, for example stabilized dune, alkali sink, or freshwater marsh. Others use systems that emphasize life forms, for example annual grassland or alpine dwarf shrub. Phillip Munz and David Keck in their *California Flora* list eleven vegetation types with twenty-nine plant communities. Plant geography professor Arthur Kuchler of the University of Kansas uses a system of nine vegetation types and fifty-five communities. The U.S. Forest Service CALVEG description has forty-two "cover units," and the California Department of Forestry and Fire Protection maps thirty-two "cover types." The California Native Plant Society has compiled a composite list of more than 200 plant community types.

Another problem is that some systems map existing vegetation, while others map pristine (prior to European contact) vegetation. It is impossible to know the actual extent and nature of California's pristine plant cover because in many places—especially at low elevations—the changes have been enormous, while in other places the changes have been subtle or only qualitative. However, good estimates can be made based on climate, soil, topography, and present-day vegetation patterns.

Spanning more than ten degrees of latitude, California is a bridge between subtropical deserts in Mexico and temperate rainforests in the Pacific Northwest. Its western border is oceanic, its eastern is continental, and in between the land surface undulates from the highest to the lowest points in the lower forty-eight

Climatic Diversity

states. The geography of California creates climatic diversity, which in turn creates vegetational diversity.

Much of California has a climate called mediterranean, similar to that of lands bordering the Mediterranean Sea. Winters are cool and wet, summers are hot and dry. Hard frost and snow in winter are rare, but winters can be depressingly damp, cool, cloudy, or foggy. Summer temperatures may be over 100°F, and one cloudless, arid summer day follows another, nearly unbroken from May through September. Almost all of the ten to thirty inches of annual rain falls in the winter months of October through March. This is not a common type of climate, and it is found in only three other parts of the world besides California and the Mediterranean rim: central Chile, the Cape region of South Africa, and southern and western parts of Australia.

Not all of California experiences a typically mediterranean climate. The northwest is too wet, the high mountains too cold, and the deserts too arid. However, more than half the state's area is mediterranean.

Along the coast there is little temperature change from day to night or from season to season. This is because the ocean is a heat sink: it releases heat to the air in winter and absorbs heat in summer. In addition, the offshore current is cold, and air passing over it is cooled. When the cold air nears warm land it forms fog banks, and the fog shields the land from direct sunlight. As a result, summer temperatures along the coast are moderate, producing a maritime version of the mediterranean climate.

Moving eastward (inland) less than seventy miles, the moderating influence of the ocean is lost and typical mediterranean conditions rule. Daily and seasonal temperature oscillations are greater. For example, Davis, in the middle of the Sacramento Valley, experiences July days with a minimum temperature of 55°F and a maximum of 100°F, for a daily spread of 45°F. Along the Bodega Bay coast, directly to the west of Davis, July days have a spread of only 7°F. Winter nights are slightly cooler in Davis than along the coast, with a higher chance of frost, so the annual range of temperature—in addition to the daily range—is also greater inland than along the coast.

Aridity increases from north to south. Crescent City, in northwestern Cal-ifornia, receives seventy-five inches of precipitation a year (over six feet of water), but San Francisco receives only twenty-five inches and San Diego just ten inches. The length of the summer dry period also increases to the south.

Both temperature and precipitation are affected by the third parameter, elevation. As air pushes eastward from the ocean, it reaches mountain ranges and is forced to rise. The air cools and some of its moisture condenses and falls as drops of rain or—if cold enough—as snow flakes and ice crystals. The rate at which heat is lost with increasing elevation depends on the amount of moisture in the air, so it is not a constant. However, on average, temperatures in California mountains fall 3°F for every 1,000 feet of elevation rise. If you know your elevation and temperature, you can confidently predict the temperature at any other elevation nearby.

Annual precipitation increases about seven inches for every 1,000 feet of elevation rise, up to about 8,000 feet elevation. Above that, most of the moisture has been wrung out of the air, and precipitation lessens. Thus, the high peaks of

High peaks of the Sierra Nevada squeeze water from Pacific storms, leaving a rainshadow over the eastern deserts (Owens Valley, Inyo Co. in the foreground).

The north-facing slope (left) of this hill supports a woodland with an oak tree overstory while the south-facing slope is covered by a dense, shrubby chaparral.

California mountains can be cold deserts. Patches of snow and ice remain on them all summer because of cold temperatures, not because of heavy precipitation. These mountain climates are too cold and wet to be called mediterranean. An upper elevational limit for mediterranean climate is about 3,000 feet in northern California and 5,000 feet in southern California.

Finally, winds from the west push over the mountain tops and air descends along their eastern flanks, heating up and drawing moisture from the soil and vegetation. Eastern California lies in an arid rainshadow cast by the mountains. It is this rainshadow which contributes to the desert climate and corresponding sparse desert scrub vegetation. Points of equal elevation on western and eastern flanks show the rainshadow effect. For example, at 7,000 feet elevation on the west flank of mountains in central California is a zone of maximum precipitation. Cathedral-like red fir (*Abies magnifica*) forests are frosted with a cumulative fall of thirty feet of snow in winter. At the same elevation on the east flank, open pine woodlands with desert shrubs beneath receive modest snowfalls and only twenty inches of rain a year. The mountain mass shadows this east face by drawing off cloud moisture well before the westerlies get this far. Even farther downslope and to the east, the California desert receives less than ten inches of rain a year. The darkest rainshadow is cast upon Death Valley, which receives an annual average of only 1.6 inches of rain.

Topography also affects climate. Slopes which face south or west are usually warmer and drier than slopes which face north or east. Northern slopes are shaded from direct sunlight, while southern slopes receive sunlight almost at right angles to the ground. Southern slopes will experience higher temperatures than northern slopes, and water will evaporate from soil and plants faster. Eastern slopes receive morning light, when temperatures are relatively low, while western slopes receive afternoon light at the warmest time of the day, so western slopes are more arid. As a result, the vegetation of opposite slopes is dramatically different, even when elevation, distance from the ocean, direction of prevailing wind, and regional climate are the same.

The Geologic Mosaic

The bedrock of California has been derived from many sources. Sediments that accumulated on the ocean floor were gathered as land by the shifting western edge of the continent. Volcanic deposits, formed from molten magma, were spewn as ash and lava onto the land surface. When magma was injected between existing layers of bedrock, it cooled slowly and crystallized as a different array of rocks and minerals. Further modifications were brought about by faulting, mountain uplift, glaciation, and erosion over long periods of time. California is now a mosaic of many kinds of bedrock, each with its own set of chemical and physical characteristics. Those characteristics directly influence soil development, which in turn influences the distribution and abundance of plant life.

Continents and oceanic basins can be compared to slabs or plates which sit on top of the mantle. They—and the outermost crust just beneath them—are not fixed. Along oceanic ridges, new, hot, nearly molten mantle material reaches the surface and exudes, building new plate material on both sides of the ridge. At places of origin, the plates move away from each other two to eight inches a year; at other places where plates collide, one plate is pushed beneath the other.

One major ridge in the earth's surface, called the eastern Pacific ridge, now stretches from the Alaskan coast, through California along the San Andreas fault, then southwest from the Sea of Cortez in an enormous arc which eventually extends south and to the west of Australia. Two hundred million years ago, several plates intersected along this ridge in an area that California now covers: the North American plate to the east; the Juan de Fuca, Farallon, Pacific, and Cocos plates to the west. A complicated period of subductions—western plates descending beneath the North American plate—built the basement material of the California land mass. With time, layer upon layer of plate material accumulated westward on the North American plate, scraped from the descending plates like so much butter by a knife.

By 170 million years ago the bases of the future Sierra Nevada and Klamath mountains were present, and by 130 million years ago the bases of the future Central Valley and much of the Coast Ranges had formed. Thirty million years ago the relative motion of the plates changed. Later the San Andreas fault system formed, splitting westernmost California from the rest of the North American plate. Land to the west of the fault moved north, relative to land to the east, at a rate of two inches per year. Granitic rocks to the west of the fault stand out as unique islands, displaced from matching rocks far to the south on the other side of the fault. The granitic headlands of Point Arena, Bodega Head, and Santa Cruz have moved approximately 300 miles north of their original placements near Taft, San Fernando, and Riverside, respectively.

More recently, uplift and volcanism have added complexity and mass to the California landscape, while erosion and glaciation have softened and suppressed the topography. A measure of this complexity is California's diversity of minerals and gems. One ecologically important rock type is serpentinite, which weathers into serpentine soil. Serpentinite is mantle material that has been brought to the surface along faults, exposed to the atmosphere, and hydrated. Its unique chemical composition—rich in magnesium, chromium, and nickel—reflects the chemistry of the underlying mantle and of the primordial whole

earth, not the modern surface crust. Its presence is a manifestation of the history of plate collisions which built California. Outcrops of serpentinite are concentrated near the San Andreas fault system. Elongated strips and ridges of serpentinite parallel the fault lines and cover 1,100 square miles of California. Very open and unusual vegetation, as well as many rare plants, grow on these outcrops. There is a greater concentration of serpentinite here than anywhere else in North America. In recognition of this, the California legislature has adopted serpentinite as the state mineral.

Terrestrial plants sink their roots into soil, obtaining moisture and essential mineral nutrients. The amount of water and nutrition available for plants in a soil depends upon the amount of clay present. Clay is the smallest of soil particles. Clay particles are microscopic and have a regular, crystal-like character, made up of atoms of aluminum, silicon, oxygen, and iron. The atoms are attached together in such a way that clay has a negative charge. As a magnet is able to attract iron filings, these bits of soil attract positively charged nutrient ions such as calcium, magnesium, and potassium. These nutrients are held by the clay in the soil, preventing them from being washed out by percolating rain water. Soils which have little clay—sandy, coarse soils—can be assumed to be low in nutrients. They are also low in water content. Water is retained as a thin film on the surface of soil particles. The finer the texture of the soil, the more surface area is available to attract and hold water.

Soil nutrients come from the raw bedrock, dust, lava, or sand that spawned the soil itself. Limestone, granite, loess, sandstone, flood-deposited silt, and a variety of volcanic, sedimentary, or metamorphic rocks all differ in chemical composition and the rate at which weathering wears them down into soil. Each parent material forms a soil capable of supporting some species, but not others. Some soils may have concentrations of particular essential elements too low for most plants to tolerate; some may have concentrations too high.

The Contributions of Soil

Granitic headlands at Point Lobos (Monterey Co.), dominated by an unusual coastal forest of Monterey Cypress and Monterey Pine.

Soil depth and slope can determine the local distribution of vegetation. Shallow or rocky soils on steep south-facing slopes support scrubby chaparral (left), while grasslands are found on gentler slopes with deeper soil (top of hill). An oak woodland is often found on cooler, north-facing slopes.

Serpentinite rock, for example, produces a soil that is low in the vital elements calcium, nitrogen, and phosphorus, but high in the heavy metals chromium and nickel. Only certain plant species can tolerate that combination of critically low essential nutrients and toxically high levels of heavy metals in soil. Biologically, serpentinite is a difficult substrate to live on. The vegetation of serpentine soil is characteristically open and scrubby. Islands of serpentine scrub lie surrounded by oceans of forest, the boundaries between scrub and forest corresponding exactly with boundaries between different parent rocks.

It generally takes several hundred to a thousand years to form each foot of soil depth. The deeper the soil, the more moisture can be stored and the more nutrients are available for root systems. Very young soils are shallow and cannot support the same plant species and the same rate of plant growth as deeper and older soils. In the Califor-

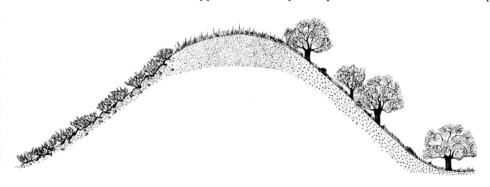

nia foothills, grassland typically sits atop the deepest soils on gentle slopes, while a dense scrub called chaparral covers the shallowest soils on steep slopes. In some places, the distance between grassland and chaparral can be traversed in two paces. Clearly the two kinds of vegetation are not separated by different climates. They occupy their different sites because of what goes on below ground.

Now imagine one rolling square mile of California. Its hillsides face in different directions. Some are steep and some are gentle. Add to that square mile a variety of soil types differing in chemistry, texture, and depth. This square mile has only one climate, but the vegetation is like a patchwork quilt. Grassland, scrub, woodland, and forest alternate with each other in a jigsaw pattern resulting from differences in topography and soil. Each patch of vegetation, then, is where it is because of local conditions—the microenvironment—and not just because of the regional environment.

Fire: Destroyer and Creator

Mediterranean climates are fire climates. In fire climates storms often create lightning strikes without rain. A certain fraction of these strikes will set fire to the landscape. Not only do many Californian plants survive fire, but some appear to require fire in order to complete their life cycle or to remain vigorous. Some vegetation types that require fire to maintain their hold on the land are completely consumed by a fire. Others have an architecture that directs fire down to the ground, where it consumes litter, herbs, shrubs, and saplings, but leaves overstory canopies unharmed.

The frequency with which fires occurred in pristine California is preserved in the growth rings of old trees. If a fire scars the outer part of a tree trunk but does not kill the tree, the scar in time will be grown over by new wood and bark and will become buried in the wood. The frequency of fires can be determined by counting the number of growth rings between scars. Some long-lived trees have survived many fires, and have many scars. Fires also leave beds of charcoal on the ground

Fire in this oak woodland (above) is consuming the understory grasses but leaving the overstory trees only slightly damaged.

A new understory of colorful herbs, such as these Chinese houses (left), often develops during the spring following a burn.

surface, and in time these become buried by new soil. In some places, layer after layer of buried charcoal occur beneath the soil surface. The age of the charcoal can be estimated by a process called carbon dating.

Analysis of these kinds of evidence shows that fire frequency was not constant over the entire state. Its frequency depended on the type of vegetation. Grasslands burned more frequently than scrub, scrub burned more frequently than some forests, and other forests burned rarely if at all. In general, fire was a regularly expected natural event in many Californian vegetation types below 6,000 feet elevation, and the same acre of ground could be expected to burn every ten to fifty years. Fire was uncommon in wetlands, deserts, and at high elevations.

We also have historical accounts, by early pioneers, as to the frequency with which they saw natural fires and the intensity with which these fires burned. John Muir witnessed both kinds of fires on a summer day while traveling through the mountain forests of the southern Sierra Nevada.

> I met a great fire, and as fire is the master scourge and controller of the distribution of trees, I stopped to watch it. . . It came racing up the steep chaparral-covered slopes of the East Fork canyon with passionate enthusiasm in a broad cataract of flames, now bending down low to feed on the green bushes, devouring acres of them at a breath, now towering high in the air as if looking abroad to choose a way, then stooping to feed again. . . But as soon as the deep forest was reached the ungovernable flood became calm like a torrent entering a lake, creeping and spreading beneath the trees where the ground was level or sloped gently, slowly nibbling the cake of compressed needles and scales with flames an inch high, rising here and there to a foot or two on dry twigs and clumps of small bushes and brome grass.

In the past 200 years, human immigrants to California have sometimes increased the frequency of wildfire and sometimes decreased it. Our present public policy of fire suppression is only eighty years old, and it follows an earlier pioneer period during which carelessness and ignorance increased the frequency and intensity of wildfires beyond those of pristine times. By increasing, then decreasing, the natural fire frequency, we have had a profound and unexpected effect on California's vegetation. In the absence of fire, plants requiring fire cannot reproduce successfully.

The Changing Landscape

Does anything ever stay the same? Seen from the perspective of geologic time, we do indeed live on a restless earth. Land masses shift, plants and animals evolve and migrate, and climates undergo enormous variation. The topography, climate, and vegetation of California today are relatively recent. A patch of ground now supporting arid desert scrub at one time held subtropical rain forest, and it may do so again in the future. These regional vegetation shifts are driven by global climate changes.

Climate and vegetation have changed dramatically during the long history of the California landscape. Daniel Axelrod, of the University of California, has spent a lifetime reconstructing the geologic history of California's vegetation. Through his work we know something about the changes in plant cover which have taken place over the past sixty million years. Three great groups of plants, called geofloras, have played a role in the region's biotic history, mixing and supplanting each other as the climate changed.

Tree ferns, palms, cycads, and large-leaved tropical plants found in the Californian geologic record represent a Neotropical-Tertiary geoflora. This flora dominated the earliest fossil records of vegetation. The presence of these plants implies a past climate with heavy summer rain and dry, frost-free winters. With later climatic cooling and drying, these plants became restricted toward the south and the coast. Today they are mainly found in southern Mexico and Central America. Fragments, such as California fan palms (*Washingtonia filifera*) in desert oases, are all that remain in modern California.

Conifers such as spruce, certain pines, fir, and deciduous hardwoods such as maple, beech, and elm in the geologic record represent an Arcto-Tertiary geoflora. These forests, of forty million years ago, were much richer in woody species than those of today. An uplift of mountains, a drop in summer rainfall, and an increase in extreme temperatures caused the rich forests to fragment. Redwoods became restricted to mild coastal climates; pines and firs moved upslope into montane forests.

A Madro-Tertiary geoflora, named for the Sierra Madre Occidentale of Mexico, is the most recent element to dominate California. It is represented in the fossil record by close relatives of modern madrone, live oak (*Quercus agrifolia*), pinyon pine (*Pinus monophylla*), chaparral species, and desert shrubs. It moved into California from Mexico as the mediterranean climate became more pronounced, about ten million years ago. Today, descendants of this geoflora dominate low-elevation vegetation throughout much of the state.

In the midst of a glacial age one million years ago, California experienced temperatures several degrees colder than at present, and desert vegetation was pushed far to the south. Many mountain features were carved at that time by glaciers which descended as low as 4,000 feet in elevation, then retreated to high alpine cirques around 12,000 years ago. About 8,000 years ago the climate was several degrees warmer than at present; desert vegetation exploded outward and timberlines pushed higher. That warm, dry period—called the Xerothermic—ended about 5,000 years ago. The climate then became cooler and wetter like that of today.

Our most recent 400 years of climate appear to have been stable, at least according to one study based on tree ring analysis. Besides telling the age of a tree and its fire history, woody plant growth rings document past climate. To read this docu-

Changes over Geologic Time

Changes in Recent Times

mentation a precision borer can be used to extract a pencil-thin tube of wood from a tree. Under a low-power microscope, rings can be measured and inspected. The wetter the growing season, the wider the growth ring.

Joel Michaelsen and others used tree rings from stands of big-cone Douglas-fir (*Pseudotsuga macrocarpa*) to reconstruct a precipitation history for central Santa Barbara County. They found little cumulative long-term change in overall mean rainfall in the last 400 years, but major fluctuations within that period. The years 1841-45 were the driest, with average rainfall of ten inches per year. The twenty-five year period 1841-64, a time of intense livestock grazing, was marked by drought. Major vegetation changes occurred at this time in the grassland valleys of California. The wettest period was 1905-09 with an annual mean of twenty-six inches.

Today there is evidence that the climate is again warming and drying, possibly because of human activity. Over the past 100 years, the concentration of carbon dioxide in the atmosphere has increased nineteen percent. Another century of such change will raise global temperature several degrees, enough to melt polar ice, raise sea level, and shift the location of land habitable for humans. Studies of the likelihood of this change occurring, a search for the first symptoms of it, and projections of what the change would do to current vegetation and land use patterns are frequently cited in the media under the general headings global warming and climate change.

The U.S. Environmental Protection Agency recently submitted a report to Congress on global warming, and one chapter of one volume dealt with the potential effects on California. The report reached several conclusions. Global warming could cause higher winter and lower spring runoff, reducing water storage in reservoirs by seven to sixteen percent, thus increasing the difficulty of meeting water supply needs. A rise in sea level would increase the area covered by water in the delta by fifteen to thirty percent, and saline water would move inland two to six miles, modifying the location of wetlands and the relative abundance of marine species. Drier conditions would reduce forest density; warmer temperatures would degrade subalpine lakes; ozone pollution would increase due to warmer temperatures; and electricity demand would escalate. The impact on agriculture is uncertain and will vary with species, but overall yields are expected to be reduced four to forty percent.

Predictions about the future vegetation of the state are highly conjectural because few experimental studies have been made of how the growth of plant species is modified by modestly warmer average temperatures. Two that have been done suggest that ponderosa pine (*Pinus ponderosa*) would increase its range and abundance in California because of its drought tolerance. Other predictions can be made from what we know of plant distribution during the Xerothermic, 5,000 to 8,000 years ago. In the mountains, we would expect increasing fire frequencies, an opening of forests so that they resemble the more arid woodlands found today

on the east slope of the Sierra Nevada. Pines and shrubs would increase in abundance, firs would decrease. Some high-elevation alpine communities and species would become endangered and even locally extinct. In contrast, the distribution of oaks and oak-dominated vegetation at lower elevations would not be expected to change significantly.

On the other hand, it could be equally likely that the pattern of glacial advance and retreat, characteristic of the past two million years, is not yet broken. It may get colder instead of warmer. We may, in that case, be in the middle of an interglacial, and a new glacial advance will begin 15,000 years in the future, bringing colder temperatures and a retreating sea level. Small comfort, however: a glacial advance will have as profound an effect on the location of habitable land as a warming climate. Either way, humans will face an enormous challenge, and California vegetation will inevitably change in response to both climate and civilization. If global warming is the issue, we may have the power to moderate the rate of change by reducing carbon dioxide emissions and increasing photosynthesis. These objectives could be met by controlling burning and car emissions and by planting vegetation.

These Douglas fir trees are slowly overtaking a woodland of oaks and bay trees during succession.

Natural Disturbance and Change: Succession and Climax

Change is also caused by local environmental catastrophes such as fire, disease, drought, avalanche, and flood. These natural disturbances recur episodically, sometimes at predictable intervals on the order of decades, sometimes randomly over centuries.

Following such a disturbance, vegetation recovers in a process called succession. Plants either reinvade the site by seeds carried in on wind or by animals, or they begin growth in place from whatever living parts remain. For example, an all-consuming wildfire will transform a montane conifer forest into ash and bare ground. In a few years there may be a meadow of perennial grasses and forbs; in a decade, a parkland of aspen (*Populus tremuloides*) growing over the herbs; in a century, an invasion of young conifers beneath the aging aspen; and finally, hundreds of years later, if left undisturbed, a return of the original conifer forest. If nothing replaces this forest it is said to be a climax type of vegetation.

Since this process of recovery can be so much longer to complete than a human life span, the reality of continuous change may escape the casual observer. Yet if changes do inexorably accumulate, then the vegetation is not yet a climax type.

Change Caused by Non-native Human Populations

Some recent changes in plant cover in California have been due to human intervention. Plants have been purposely or accidentally brought to the state from other parts of the world since the time of the first non-native settlers. In 1769 Father Junipero Serra founded the first Spanish settlement in Alta California at San Diego Bay. He brought at least three exotic weed species with him. By 1824 at least sixteen exotic species had become established in California. We know their number and their date of introduction because identifiable plant parts were trapped in adobe bricks made to build the chain of missions. When botanists break open the bricks, they can deduce the existence of some of the past surrounding plant life by sorting through plant parts accidentally baked into the clay.

During the Mexican occupation of 1825-48, an additional sixty-three species were added. Pioneer American settlements of 1849-60 brought fifty-five more. In 1925, 292 species of non-native plants were part of the flora; in 1951 there were 437; in 1959, 797 species; in 1968, 975 species, and in 1993, a total of 1023 species.

Nearly a thousand species is a large number, but we can be sure that many times more species arrived here (and continue to arrive) but were not successful in establishing themselves. In addition, many garden and crop plants, ornamental shrubs and trees, and wild plants have been brought to California that cannot live and reproduce outside cultivated landscapes. There was something unique in the biology of those one thousand successful weeds that permitted them to run wild in California's landscapes. What made these plants successful?

Changes in humidity cause the corkscrew-shaped fruits of weedy filaree to drill themselves into the soil for better germination.

Herbert Baker, a botanist at the University of California in Berkeley, once concocted a list of traits that the ideal weed would possess. No single weed has all of those traits or we would be ankle-deep in that species. Baker's list of traits for a successful weed includes fast growth rate, early maturity and reproduction, an abundant production of seeds, a prostrate habit, fragile stems, the capacity for wind- or self-pollination, tolerance of full sun, and seeds capable of long-term dormancy in soil. These traits are most easily combined in small, annual, herbaceous plants. Large, perennial, woody weeds such as gorse (*Ulex europaeus*), eucalyptus, and Scotch broom (*Cytisus scoparius*) are obvious in the landscape, but their numbers and the territory they have claimed are modest compared to annual herbs. Most California weeds originated in other regions with mediterranean climates and occupied disturbed or degraded habitats there, so they were well adapted to the California of the nineteenth and twentieth centuries. We may be sure that other potentially successful weeds still remain to be introduced to and established in California.

Other types of human activity are changing the environment, and these may change the landscape too. Smog is formed, cattle are allowed to graze, ozone is depleted, and natural fires are put out. These are disturbances which set a succession in motion. What had been climax vegetation for millennia now may be successional, and a new climax vegetation will be established. These future climaxes are not predictable. Under the stresses of ozone and acid deposition, what will the new climax montane conifer forests become? With continued fire suppression, what new climax will chaparral become? What new desert scrub climax waits for us, as a result of overgrazing?

These stresses have been caused by a rising human population, as well as a shift in its ethnicity and cultural habits. Two hundred years ago there were an estimated 300,000 Native Californians in California. Today, there are nearly thirty million humans, largely of European, Asian, African, and Latin American extraction. Fewer than 50,000 Native Californians remain. Present human demands on the environment are drastically different from those of the past. Traditional technologies of original California peoples were based on the philosophy of resource conservation and renewal. Use of their sophisticated technologies meant modest impact on the ecosystem.

Curly dock, one of the first weed species to reach California during Spanish settlement. The small flowers to the right are of another exotic species, chicory.

In contrast, European technologies are based on an agricultural economy that changes wildness to a homogeneous, managed landscape. The diversity of native vegetation, animals, and cultures is threatened by this philosophy and practice, but the notion of endless prosperity based on natural diversity is still part of the California myth.

Join us—in the remaining chapters—on a journey through space and time as we look at California's cover from coast to desert, from past to future, from pristine to managed, from beautiful to degraded, from known to imagined. Our journey will emphasize the structure and diversity of California vegetation and the processes that create and maintain it. We will also examine the nature and consequences of human-induced change and the need for conserving this vital resource.

2

THE COASTAL INTERFACE

ALIFORNIA'S WESTERN EDGE IS A RESTLESS INTERFACE BE-
tween land and sea. It is a place where frayed but defiant rocks of the
North American continent are continually assaulted by powerful
forces of the Pacific Ocean. As a result, landforms along the 1,100-mile
coastline are as spectacular as they are diverse: graceful sandy bays, muddy tidal
flats, precipitous cliffs, rocky coves, and fluent, undulating dunes. Each landform
differs with respect to geology, soil, exposure, and other ecological characteristics
of importance to plants. The landscape patterns of coastal vegetation result from
those forces and forms at the land-sea interface.

The dramatic topography of the California coast reflects the strength and
persistence of the ocean's mechanical forces. The pull of moon gravity and the
push of onshore wind produce vast amounts of tidal and wave energy that is
unleashed against the continental edge. The crush of a storm wave, inundation
by seawater, and the blast of the wind continue to erode seacliffs and deposit beach
sand. These processes determine much about the vegetation of coastal lands. Here
plants tend to be creeping, rather than erect, short rather than tall, and flexible
rather than rigid, thereby minimizing exposure to the mechanical forces of wind
and wave.

The ocean also exerts chemical forces over the land and its vegetation.
Each year tons of water-borne minerals are brought onshore by tides, wind, fog,
and storm waves. Sea water is approximately 3.4 percent dissolved minerals,
including sodium chloride, calcium sulphate, potassium nitrate, magnesium
carbonate, iron phosphate, and many others. Some minerals, such as nitrate,
potassium, and phosphate, are essential to the metabolism of coastal plants rooted
in sandy, nutrient-poor soils. These are used to make proteins, molecules of the
genetic material DNA and RNA, and living machinery for trapping the sun's
energy during photosynthesis. Sodium and chloride (together comprising table
salt), however, are deposited in such quantities that they may severely retard
plant growth. As soil salinities rise above 0.5 percent, plant metabolism becomes deranged in all but the most salt-tolerant species. Sea water exceeds this
threshold, so when salt is carried onto the land by wind, waves, or high tides,

*Douglas Iris, a showy
perennial of coastal prairie.*

it becomes a significant stress factor in the lives of plants along the coastal interface.

Salts become stressful at high concentrations for several reasons, but mainly because water moves across semi-permeable cell membranes from less salty areas to more salty areas. For plants without special adaptations and surrounded by seawater, the direction of flow would be out of the root, instead of into it; as a result, the plant would wilt and ultimately die. High salinity affects the roots as though the soil were dry—a physiological drought.

Salt-tolerant species are able to keep water flowing into the plant by several techniques. Some species, such as perennial pickleweed (*Salicornia virginica*), accumulate salt in their tissues to concentrations exceeding those in the saline soil. The net effect is to reestablish normal water flow from the soil into the root. These plants generally store excess salt in turgid, succulent tissue. The cells of succulent tissue are enlarged, swollen by a membrane-bound sac called a vacuole. Most of the cell's volume is occupied by the vacuole, leaving the living cytoplasm pressed to the outside against the cell wall. The vacuole does not contain pure water, but salt water. Salts are taken up by roots from the soil solution, moved in the plant's internal water stream to above-ground parts, and then shunted into cell vacuoles. Succulents tolerate salt, then, by accumulating it away from the metabolically active cytoplasm.

Saltgrass (*Distichlis spicata*), in contrast, excretes excess salt as a concentrated liquid which dries on leaf surfaces into glistening crystals. Scurfy saltbush (*Atriplex leucophylla*) deposits salt in hair-like glands on the leaf surface.

Coastal vegetation is composed of a salt-tolerant subset of California's flora which have these adaptations. But even such species have tolerance limits that can be exceeded during exceptional storm events. When salt spray and waves are carried farther inland than usual or for a longer time or more intensively, salt-tolerant plants may be killed. That portion of the coast will then be barren of vegetation until colonized by inland survivors or oceanic castaways.

Beach and Dune Communities: the Violent Interface

Beaches and dunes are created where great quantities of sand are deposited onshore by waves and transported inland by winds. The sand is carried to the ocean by free-flowing rivers that erode the interior of the state. Sand bars and spits form at river mouths only to be shattered by waves and released to the ocean's currents for redistribution along the coastline. Pacific waves thrust the ocean's sand onto the shore, where it is distributed by wind as a veneer over the mainland bedrock.

Once ashore, sand becomes available for plant colonization above the intertidal zone. Winter storm waves, however, often overwash the beach, reclaiming great quantities of sand and obliterating entire populations of pioneering beach species. Overwash, salt spray, and tumbling seafoam introduce bursts of salt to this severe interface. The dramatic cycle of beach accretion and erosion,

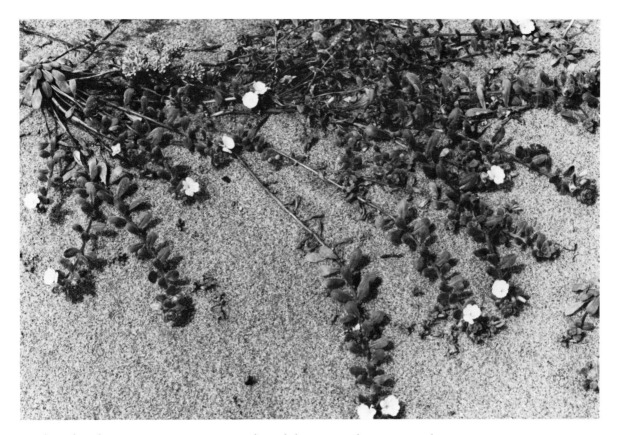

combined with exposure to oceanic swash and drying winds carrying salt spray, create harsh conditions for the development of beach vegetation.

California beach vegetation extends from mean high tideline to the foot of the first dune, or foredune. It consists of low-growing, open and trailing plants that cover only ten to twenty percent of the ground. Beach vegetation seldom amounts to more than a single canopy layer consisting of only seven or eight species at any one location. Near the water, where driftwood and kelp come to rest in limp, tangled heaps, there is a barren zone where plants can't survive the chaos of wind and wave. Low beach lies just above this zone, and its sparse vegetation is composed of an introduced species, a sprawling annual plant called sea rocket (*Cakile maritima*), and a native perennial species, scurfy saltbush.

Inland from this zone of scattered sea rockets, gentle sand hummocks are spattered with low-growing herbaceous perennials, including some stout beach grasses. The lee of these hummocks, where salt spray is reduced, provides the only shelter for species with lower salt tolerance, such as silver beachweed (*Ambrosia chamissonis*). Given the harsh physical and chemical environment, it is not surprising that beach species at the leading edge of vegetation share many growth form characteristics: creeping rhizomes or stolons allowing rapid recolonization of

Low growing species, such as beach evening primrose, usually dominate beach and dune vegetation where exposure to salt spray and sand blast is greatest.

new beach sand; prostrate habit keeping plants away from sand blast and salt spray; succulent leaves which can store harmful salts; and a covering of fine, white hairs that reflect excessive amounts of solar energy in this open, exposed habitat.

Adaptations to Stress in Beach Habitats

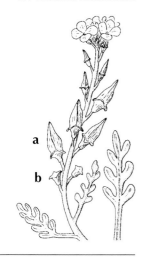

Searocket fruits (above): one breaks off (a) and one stays attached (b) to the parent plant.

A cross-section (right) beginning at the ocean (a) and extending across the tideline (b) with berm and deposits of wrack. Just beyond is the low beach (c), nearly devoid of plants because of frequent storm waves. Further inland, beach plants increase in abundance, peaking on the foredune (d). Wind-blown depressions, or swales (e), are wetlands. Stabilized dunes (f) are dominated by woody shrubs and coastal forests (g) are found inland where there is little or no salt spray.

Plants that pioneer the low beach possess special adaptations for dealing with stresses at the violent interface. Sea rocket, for example, has floating fruits that wash ashore with the ocean's assorted debris. It grows rapidly, flowering and setting fruit early in life. This rapid maturation takes advantage of the brief intervals between major storms when waves sweep the low beach and leave behind a polished sand surface. Storm waves bring rebirth as well as death, burying ripe fruits in place and carrying others out to sea and onto distant beaches.

Sea rocket has two kinds of fruit, which allows the plant to hedge its bets on where the best habitat will be the next growing season. Although the parent plant may have thrived where it lived the current year, the strand is a dynamic place and that spot may or may not be favorable to growth next year. Sea rocket splits its seed bank, putting half near the parent and sending the other half out to sea. Each fruit is a two-parted corky container shaped like a miniature rocket about one inch long. Many are produced on each long flowering branch. The top half of each fruit, containing one seed, easily breaks off from the bottom half and can be carried by waves and offshore currents to other beaches. As long as the fruit wall is saturated with salt water, the seed remains dormant; but once deposited back on land and leached by rainwater, it will germinate. This represents the fate of half the seed crop. The other half stays put. The bottom half of each fruit remains firmly attached to the dead or dying parent, and is buried in place by moving sand. Next year, if the spot is still habitable, hundreds of seedlings will germinate there and some will survive to replace the parent plant.

Scurfy saltbush is a pioneer plant with different adaptations for dealing with life along the coast. This plant is closely related to the desert saltbushes of Baja California and the Colorado River lowlands (see Chapter 6). It has glands and hairs that excrete salt to the outside surface of leaves and stems, leaving it as a crust of harmless crystals. Scurfy saltbush also carries out a specialized kind of photosynthesis called C_4 (pronounced c-four) photosynthesis.

Photosynthesis is the absorption of carbon dioxide from the air and its combination with water within a green cell to produce sugar, a carbohydrate.

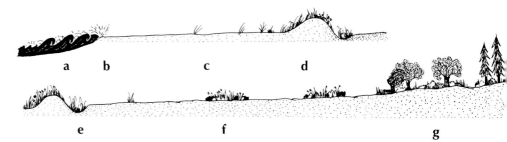

Photosynthesis is driven by the activity of some important plant enzymes, as well as by side reactions which convert sunlight into chemical (caloric) energy and return oxygen to the atmosphere. The absorption and transformation of carbon dioxide is called carbon fixation. Carbon dioxide enters leaves through microscopic pores (stomates) in the surface. Typically, the pores open during the day and close at night.

Most plants fix carbon and produce sugar through a series of biochemical steps called C_3 photosynthesis. In the 1950s, Melvin Calvin, then at the University of California at Berkeley, received a Nobel Prize for being the first to describe these steps. An early step in the pathway results in the formation of a carbohydrate with three carbons, hence the name C_3 photosynthesis. Sugar, which has six carbons, is formed as a last step in the pathway.

For many years, it was believed that all plants fixed carbon this way, but now it is known that a cluster of species in certain plant families fix carbon by a different path. An early product is an organic acid with four carbons, hence the name C_4 photosynthesis. Sugar is ultimately produced by this pathway, but the steps involved and certain key enzymes are different from those in C_3 photosynthesis. There are ecological differences as well: C_4 photosynthesis confers on plants that follow this metabolic pathway greater tolerances of salinity, high temperatures, intense sunlight, and drought. Water use is more efficient, less water being used for each unit of sugar created.

Consequently, C_4 species are often found in habitats that are saline, hot, open, and dry. Most habitats, however, have both C_3 and C_4 plants living side by side. The beach is no different: sea rocket is a C_3 species and scurfy saltbush is a C_4 species.

Large, fleshy taproot of yellow sand verbena (top), a frequent member of high beach vegetation.

Succulent, erect leaves of yellow sand verbena (bottom).

High beach is only three or four feet above low beach, but this is enough to diminish storm wave overwash and permit colonization by additional plant species. Prominently inhabiting the high beach are two types of sand verbena, yellow (*Abronia latifolia*) and purple (*Abronia maritima*), commonly found north and south of Morro Bay, respectively. These natives have long, scrambling stems that emanate from the top of a fleshy, deep taproot and spread across the hummocky beach surface. Both have thick, succulent leaves that orient themselves vertically, perhaps to minimize contact with hot sand and to reduce solar exposure. Several other species, including beach morningglory (*Calystegia soldanella*), coast strawberry (*Fragaria chiloensis*), and beach evening primrose (*Camissonia cheiranthifolia*), also have thin trailing stems and thick leaves. The sand surface of the high beach often looks sewn together by a sparse growth of intertwining threads, decorated with green, leafy sequins. Other high beach species with broad leaves and trailing stems include silver beachweed, succulent sea fig (*Carpobrotus aequilaterus*), and beach pea (*Lathyrus japonicus*).

The Grass-roots of Beach and Dune Stabilization

Sticking up between the threads and sequins of broadleaved plants on the high beach are clumps of beachgrass. Most of the grass is exotic European beachgrass (*Ammophila arenaria*), first introduced to North America in 1869 and now the most abundant beach and dune plant along the California coast. It was widely planted as a sand stabilizer to prevent the filling of shallow harbors and burial of coastal roads and railroad tracks by windblown sand. Golden Gate Park in San Francisco was created from sand dunes stabilized by this grass. It has subsequently spread on its own, because waves and currents are able to disperse pieces of living rhizome.

European beachgrass forms dense swards of erect leaves and stems, especially in sheltered or elevated portions of the shore where storm waves are rare. There is little room for native plants in such a dark, uniform stand, so a beach community dominated by European beachgrass has far

A dune dominated by dense swards of European beachgrass, introduced to the Pacific coast during the last century.

fewer species of plants and insects than one dominated by native Pacific beachgrass (*Elymus mollis*). Pacific beachgrass is a stout relative of our agricultural rye grasses. It spreads horizontally by means of numerous underground rhizomes, thereby stabilizing beach sand in the face of wave erosion. Its widely spaced clumps allow plenty of room for other beach plants, so a community dominated by Pacific beachgrass is diverse in species composition and open in vegetation architecture.

This openness permitted European beachgrass to become established on the strand. From the Big Sur coast to northern Oregon, European beachgrass has supplanted Pacific beachgrass by its aggressive, competitive growth. The only extensive remnant stands of Pacific beachgrass left in California occur on Point Reyes National Seashore and Lanphere-Christensen Dunes Reserve near Arcata. At the latter location, efforts are being made to control and eradicate the introduced grass and to return the scarce high beach habitat to native plants and animals.

Above the beach, where storm waves seldom reach, unstabilized beach sand is blown into sculpted mounds called dunes. Here sand particles move at high velocity and the open dune surface undulates under the force of constant wind. Plants are subjected to sand blast, burial, and salt spray, especially on the windward side of the foredune. As a result, foredune vegetation is composed mostly of the same species found just above the tide line. The plants are low, patchy, perennial herbs and grasses with creeping rhizomes. Plant cover is usually higher than on the beach, typically twenty to fifty percent, due to the lack of destructive erosion by waves. Behind the foredune, clumps of vegetation are stationary islands in an ocean of moving sand. Douglas blue grass (*Poa douglasii*), beach pea, and beach sagewort (*Artemisia pycnocephala*) provide sparse cover.

European beachgrass has had a drastic effect on foredunes and dune vegetation. Low parabolic dunes, typical of the Pacific coast prior to 1850, were gradually replaced by high, steep dune ridges as they became colonized by European beachgrass. The rhizomes of this introduced species tend to grow vertically rather than horizontally, effectively stabilizing windblown sand and resulting in tall, immobile dunes. As the dense cover of European beachgrass develops, it prevents invasion by low-growing perennial plants and hoards sand that would otherwise rejuvenate beaches and dunes. The Pacific coastline has been radically transformed by introduction of this plant, which has contributed greatly to a loss of native plant populations and a disruption of the geologic processes that maintain windswept, unstabilized dunes.

Two other introduced dune perennials have had an impact on natural and artificial habitats. Both are prostrate succulents with large, attractive flowers: sea fig from Chile and hottentot fig (*Carpobrotus edulis*) from South Africa. These plants spread their heavy stems radially out over the sand surface, forming such dense mats that other species are crowded out. This ability to form solid cover led the California Department of Transportation to select hottentot fig as a ground cover along freeway banks and interchanges. The smaller-leaved sea fig has spread on its own from the dunes to coastal bluffs, where it often creates a single-species landscape.

A dune dominated by open stands of the native Pacific beachgrass (Lanphere-Christensen Dunes, Humbolt Co.), with clumps of searocket to the right.

Bush lupine, a native, woody stabilizer of coastal dunes.

Native plants also stabilize dunes, but they do not do so as aggressively. Woody shrubs, such as bush lupine (*Lupinus arboreus*), lizardtail (*Eriophyllum staechadifolium*), coast buckwheat (*Eriogonum latifolium*), and dune heather (*Haplopappus ericoides*) provide a two- to four-foot overstory with forty to eighty percent cover. A low understory of succulent live-forever (*Dudleya* species) and introduced sea fig is a common feature of stabilized dunes, especially from Bodega Bay to Point Conception.

Mixed with the succulent layer are species found on unstabilized dunes and beach, including Douglas bluegrass, beach sagewort, sand verbena, and coast strawberry. In addition, a number of species from adjacent non-dune communities can invade less exposed portions of stabilized dunes. These include California poppy (*Eschscholzia californica*), coyotebrush (*Baccharis pilularis*), and deerweed (*Lotus scoparius*). The effects of high wind, storm waves, salt spray, and sand

movement are greatly diminished in stabilized dunes, and even more so farther inland
where dune vegetation gives way to coastal
prairie, scrub, and forest communities.

Among the dunes are deep depressions,
or swales, where wind erosion has completely
removed the sand veneer down to the water
table. Swale vegetation, in contrast to the
surrounding beach and dune, develops under
stable, moist, and salt-free conditions. Often
present are white alder (*Alnus rhombifolia*),
arroyo willow (*Salix lasiolepis*), brownheaded
rush (*Juncus phaeocephalus*), and miner's
monkeyflower (*Mimulus guttatus*)—species
commonly associated with coast forest, riparian forest, freshwater marsh, and moist mead-

ows. In time the vegetation of a swale may continue to develop into an outpost
of coastal scrub or forest or it may be buried by unstabilized sand as surrounding
dunes are shifted by wind storms.

People are often surprised to learn that beaches and dunes have natural vegetation. The typical seaside landscape in populated areas is one of barren sandy flats,
an occasional lifeguard station, and hot asphalt parking lots. However, even
beaches and dunes in the vicinity of San Francisco, Los Angeles, and San Diego
had well developed native vegetation as late as the 1930s. Simple, herbaceous
plants of the native dunes, though tolerant of salt spray, crashing waves, and
relentless wind, could not withstand daily trampling by thousands of humans. It
is still possible to see an occasional beach primrose or sea rocket in Pacifica,
Venice, or La Jolla, but most vegetation has been completely obliterated.
Residential and commercial development, sand mining, road construction, off-
road vehicle recreation, and in some cases even livestock grazing have been the
principal causes of destruction.

Once destroyed, can beach and dune vegetation be restored? Several large-
scale projects have attempted to recreate native beach and dune vegetation, with
mixed success. At Marina State Beach along Monterey Bay, large areas of
unstabilized dunes had been denuded by off-road driving, hang gliding, and other
intensive forms of recreation. Exotic species such as hottentot fig dominated the
remaining vegetation. In 1985 the California Department of Parks and Recreation began a restoration effort aimed at stabilizing barren dunes with native
species and enlarging a decimated population of Menzies' wallflower (*Erysimum
menziesii*), a highly endangered coastal plant.

Forty-three acres were fenced, hand weeded, and recontoured to prepare the

*Beach and Dune
Restoration*

*Sea fig, an introduced
perennial that stabilizes
dunes and crowds out
native species.*

Restoration of beach and dune vegetation often begins with removal of exotic species, such as European beachgrass.

site for revegetation. A total of 2,800 pounds of seeds from nineteen native species were collected at local beaches and dunes, thus ensuring the genetic similarity between new and old populations. Pioneer beach plants such as yellow sand verbena, beach evening primrose, and silver beachweed were sown in the foredune zone. Dune species such as beach sagewort, dune heather, and lizardtail were sown behind the foredune in areas where stabilized dune vegetation would naturally occur. An additional 8,500 seedlings of the endangered wallflower were grown in a nursery and transplanted among members of the existing population.

More than eighty percent of the wallflower transplants survived the first year, moving this species a little farther from the precipice of extinction. At the same time, it was found that the use of native species, without irrigation, produced adequate cover to stabilize the dunes and to provide a natural nursery for longer-lived dune species that became established over the next several years.

Success was hard won: more than 6,500 hours of labor and $250,000 in materials were required to reclaim a mere forty-three acres of the park. Native beach and dune vegetation can be repaired, but the high cost of mitigation argues against destructive uses of this scarce resource. Far better is conservation of dune vegetation before it is destroyed by thoughtless activity. Trampling, for example, can be minimized by the construction of elevated boardwalks that take people through the sensitive foredune and upper beach to the strand. Natural processes, such as blowing sand, as well as the plants themselves, are unimpeded by such walkways. Boardwalks are uncommon along the California coast, but they have been used extensively in states such as Florida to protect sensitive dune vegetation.

The most serious threats to beaches and dunes, however, cannot be easily observed or mitigated. Damming of major rivers limits the supply of new sand needed to rejuvenate eroded beaches and stabilized dunes. Sediments once transported to beaches now collect in the bottoms of reservoirs. The link between dam building and beach preservation is masked because so many recreational beaches are intensively groomed with sand trucked from the interior of the state. The few remaining wild beaches, such as those of Humboldt Bay, Point Reyes, and Monterey Bay, do not receive sand supplements. Threats such as sand starvation are being recognized by park managers, coastal zone planners, developers, and naturalists. The strong public mandate that already exists for coastal access needs to be expanded to protect the natural processes that maintain beach and dune systems.

Coastal Prairie and Scrub: Away from the Interface

The modern topography of California's coast has been molded by colossal forces such as colliding continental plates, crustal heaving and uplifting, and changes in sea level. Cliffs, terraces, and rolling hills along the coastal zone are the result of these forces. Their surfaces are away from surf and moving sand, and somewhat removed from the stresses unique to beaches and dunes. Consequently, they are vegetated with different species, different growth forms, and different plant communities. We walk here, on foggy summer days, through coastal prairie and coastal scrub.

Coastal Prairie

North of Big Sur the inland edge of stabilized dune scrub blends into a special form of grassland called coastal prairie. The maritime climate is cool and moist, but the harsh physical and chemical conditions found near the beach are absent. Uplifted marine terraces provide parent material for the development of productive, deep, moist, well drained grassland soils. On these terrace soils is a rich assortment of perennial grasses and showy broadleaved herbs, some of which extend well north of California. Oregon hairgrass (*Deschampsia caespitosa* ssp. *holciformis*), Idaho fescue (*Festuca idahoensis*), California oatgrass (*Danthonia californica*), and as many as twelve other grasses can form a two- to three-foot canopy of leaves, stems, and floral stalks that wave under autumn winds and glisten with droplets of summer fog.

Blue wild rye.

These native perennials are often described as bunchgrasses because they do not have spreading rhizomes and they tend to grow in clumps. Erect grass stems emerge from a single, discrete rootstock, allowing individual grass plants to be easily identified. Plants are often separated by patches of other species or by open ground. Herbaceous species such as Douglas iris (*Iris douglasiana*), goldfields (*Lasthenia* spp.), California buttercup (*Ranunculus californicus*), baby blue-eyes (*Nemophila menziesii*), and tidy tips (*Layia platyglossa* ssp. *platyglossa*) produce bright floral displays between the bunchgrasses in late April and May. On rocky outcrops with no soil, staghorn lichens comb water from the fog and succulent live-forevers huddle in damp crevices. Where soils are shallow and dry, coastal prairie resembles the central valley grasslands, with purple needlegrass (*Stipa pulchra*), pine bluegrass (*Poa scabrella*), and annual grasses and herbs (see Chapter 4).

Where coastal terraces have been thrust upward, exposing sheer walls of barren sandstone and granite, the seaward edge of prairie is influenced by the ocean's wind and salt. Here a narrow strip of grassland adjacent to the precipice is replaced by a unique bluff edge community. Sea-pink (*Armeria maritima* var. *californica*), lizardtail, beach sagewort, coast buckwheat, and seaside daisy (*Erigeron glaucus*) form a low, tangled, wind-beaten, salt-laden canopy. The shallow soil and intense salt spray prevent establishment and growth of perennial grassland plants. Winter rains dilute the effects of wind-borne salt and allow annuals such as California poppy to temporarily invade the bluff edge area in spring. As the rains diminish these annuals bloom, set seed, and die back in a wave that recedes from the bluff edge.

An early pioneer described the lower Russian River valley (Sonoma County) as "a great wide area of waving grasses higher than a man's head with deer, bear and other big game everywhere." Grazing by native mammals was light and

Annual plants, such as tidy tips, California buttercups and goldfields, are found in coastal prairie during the spring (Pt. Reyes National Seashore, Marin Co.).

seasonal. Native people in this region, including those of the Yurok, Mattole, Pomo, Coast Miwok, and Costanoan groups, all used wildfire to increase the productivity of these grasslands and their wildlife. This practice favored perennial grasses and herbs that could quickly recover via underground buds protected by an insulating layer of soil. Woody plants with above-ground, exposed buds were discouraged by frequent fires.

Arrival of the Europeans in the middle of the nineteenth century brought fire suppression and heavy, year-round grazing by cattle and sheep. The coastal prairie in many places was transformed into a grassland of introduced species, including nearly pure stands of non-native velvet grass (*Holcus lanatus*), sweet vernal grass (*Anthoxanthum odoratum*), soft chess (*Bromus mollis*), Italian ryegrass (*Lolium multiflorum*), wild oats (*Avena fatua*), and barley (*Hordeum* spp.). Weedy herbs, particularly those unpalatable to cattle, spread indiscriminately in the absence of fire. Introduced milk thistle (*Silybum marianum*), wild artichoke (*Cynara scolymus*), and Klamath weed (*Hypericum perforatum*) came to dominate millions of acres of former coastal prairie, drastically reducing the biological and agricultural value of the land.

Klamath weed contains a toxin that causes sores on unpigmented skin, such as the skin around the muzzle of livestock. The sores lead to weight loss or death because the animal is unable to feed. In 1946 a pioneering effort in biological control was begun in Humboldt County to control Klamath weed by releasing a population of Klamath weed beetles that feed exclusively on this plant. Within two years Klamath weed had been effectively eliminated from large tracts of grassland, allowing perennial and annual grasses to flourish. Studies in other areas of the northwest have shown that careful management using fire, light grazing, and biological weed control can restore native grasses in as little as ten years. These results provide hope for the return of these once impressive prairies.

Coastal Scrub

Several shrub-dominated communities are adjacent to stabilized dune and prairie, and they extend inland over the low coastal mountains. Coastal scrub vegetation is dominated by evergreen species north of Big Sur and by drought-deciduous species to the south. In the absence of frequent fire or grazing, scrub will invade the margins of coastal grasslands and begin a chain of community succession leading to chaparral, evergreen forest, or oak woodland. The result is the swirling mosaic of grassland, scrub, chaparral, woodland, and forest characteristic of the steep hillsides and shallow valleys in coastal California.

Northern coastal scrub extends from Big Sur north into Oregon. It is a dense, two-storied assemblage of shrubs, vines, herbs, and grasses. The dominant plant is coyotebrush, a three- to six-foot-tall shrub with stiff, bright green foliage and small white flowers. The climate of the north coast, moderated by wet winters and mild, foggy summers, allows the leaves of coyotebrush to continue photosynthesis all year long. Often entwined within its canopy are the stems of poison oak (*Toxicodendron diversilobum*). Salal (*Gaultheria shallon*), California coffeeberry (*Rhamnus californica*), cow parsnip (*Heracleum lanatum*), and one or more species of shrubby lupine are also present. Bush lupine, with green foliage and yellow flowers, is widespread along the North Coast and can form nearly pure stands. Chamisso lupine (*Lupinus chamissonis*), with silver foliage and blue flowers, is found only south of Sonoma County. Bush monkeyflower (*Diplacus aurantiacus* syn. *Mimulus aurantiacus*) wedges its resinous leaves and salmon-colored flowers between other overstory shrubs. A dense understory of grasses and herbs leaves no dry trail on a foggy day.

When northern coastal scrub is disturbed by natural forces such as lightning, fire, and landslide or by human forces such as road construction, new habitat is created for the inva-sion of introduced plants. Pampas grass (*Cortaderia jubata*), a native of South America, has readily invaded north coastal scrub. This robust grass can form mounds of foliage five to six feet tall; its whitish, silken flowering stalks are ad-mired by interior deco-rators and gardeners for their soft, wind-combed appearance. But these same flower-ing stalks produce thousands of seeds that germinate and colo-nize barren soil. Once established, pampas

grass is difficult to control and nearly impossible to eradicate. Plants must be removed by hand before they are large enough to begin flowering. The failure to do so along the central coast near Monterey, Big Sur, and Cambria has been nothing short of disastrous. The once beautiful hillside blanket of coyotebrush and bush monkeyflower is now frequently interrupted and many times completely replaced by coarse tufts of this weedy grass.

Southern coastal scrub, extending from Big Sur south into Baja California, is a simpler, more open community, dominated by low-growing, drought-deciduous shrubs only two to three feet tall. Leaves are broad or narrow, soft to the touch, and capable of high rates of photosynthesis when soil water is available. Common native dominants in this community include black sage (*Salvia mellifera*), white sage (*Salvia apiana*), California sagebrush, California buckwheat (*Eriogonum fasciculatum*), and golden yarrow (*Eriophyllum confertiflorum*). In contrast to their drab, gray- green foliage, these species frequently have showy flowers in bright blues, yellows, and reds that attract a wide variety of pollinating insects.

Farther to the south succulent species become an important part of the cover:

An overstory of evergreen and drought-deciduous shrubs, three to six feet high, is typical of coastal scrub vegetation (Morro Bay, San Luis Obispo Co).

Flowers of purple sage, a drought-deciduous shrub of southern coastal scrub.

century plant (*Agave* spp.), striped live-forever (*Dudleya variegata*), and several kinds of cacti. Beyond the Mexican border this mixture of shrubs and succulents blends with species from the Sonoran desert, forming the exceptionally rich and unusual coastal vegetation of Baja California.

The soft leaves of black sage are characteristic of many southern coastal scrub species. The leaves are capable of absorbing large quantities of carbon dioxide because their metabolic resources favor photosynthesis over water conservation. Black sage leaves are rich in carbon-fixing (photosynthetic) enzymes but lack a thick, waxy cuticle and tightly controlled stomata. Photosynthesis is confined to a few wet months of the year when there is ample water in the root zone. Consequently, black sage grows rapidly from November to March and produces numerous flowers by April. During hot, dry summers most leaves are shed because they have no structural mechanism for conserving water or any physiological mechanism for tolerating dehydration. By August, only a few leaves remain attached to many barren branches.

Like other members of this community, the leaves of black sage contain piquant, distinctive chemicals that evaporate and float like fragrant musk on the warm air. The chemicals found in California's native sages and sagebrushes are similar to those found in thyme, cooking sage, and rosemary. They may have value to the shrubs that produce them by deterring foliage-eating insects or by retarding the growth of competing plants. They also evoke strong memories of coastal mountains to those who have grown up in southern California.

Memories of southern coastal scrub may be all that is left in a few decades. The gentle hills and terraces of this community are readily converted into housing and commercial developments. A major thrust of development is rapidly changing the beautiful landscapes of southern coastal scrub into monotonous seacoast suburbs. As the native scrub disappears, so will many species of plants and animals that are unique to this Californian community.

Rivers that cut through and drain coastal mountains not only transport sediments to the sea, but also allow the sea to reach inland along bays, river deltas, and estuaries. In such places the land may be alternately wet and dry, depending on river flows, tidal heights, and small differences in elevation. The water is alternately fresh and saline, depending on the mix of river and ocean and how much evaporation has occurred in the shallows. Lacking the force of ocean waves, water laps at the shore, quieted at high tide by dense marsh vegetation that emerges from the soft, muddy bottom. Lacking the force of ocean wind, the air contains only small amounts of salt spray to mist over the marsh vegetation. These places, known as coastal wetlands, form a gentle interface between land and sea.

Coastal Wetlands: the Gentle Interface

The term wetland does not stand for a single vegetation type or a single plant community. It is now also a legal term that applies to any habitat where the soil is saturated with water within eighteen inches of the surface for a period of at least one week a year. In 1991 federal agencies were debating a redefinition. One proposed redefinition would require that the soil be saturated with water to the surface for a period of at least two or three weeks a year. This would greatly reduce the extent of legal wetland habitat, permitting more widespread development.

However the depth and duration of saturation are defined, the fact is that wetlands are periodically waterlogged and plants growing there must be tolerant of low levels of soil oxygen. Only a small subset of California's flora is flood-tolerant, that is, tolerant of oxygen starvation in the root zone. The presence of flood-tolerant species is a good indication that the local site is a wetland even if the ground appears to be dry for most of the year.

Wetlands occur throughout California, and they are discussed in every chapter of this book. Wetlands include seagrass beds, coastal salt marshes (also called tidal marshes), interior salt marshes, brackish water marshes, freshwater marshes, vernal pools, riparian forests, montane wet meadows, and desert oases and playas.

Coastal wetland vegetation really begins underwater, below the intertidal zone and offshore from beaches and marshes. It is dominated by flowering perennial plants generally known as seagrasses. Seagrasses, such as the common eel-grass

Cross-section through tidal marsh, showing: a) beds of eelgrass and algae below all but the lowest tides; b) mudflat with algal crust; c) lower marsh dominated by cordgrass; d) mid-marsh dominated by perennial pickleweed; e) upper marsh with a mix of tall pickleweed, salt grass, and alkali heath; and f) upland vegetation on non-saline soil above the intertidal zone (not drawn to scale).

Seagrass Beds

Cordgrass in low marsh, with perennial pickleweed and gumplant (shrub with tall flowers) in mid-marsh (China Camp State Park, Marin Co.).

(*Zostera marina*), are capable of growing completely submerged in sea water for their entire life span. Extensive beds of waving eel-grass occupy shallow bays, lagoons, and harbors protected from heavy seas by enveloping shorelines. In these quiet waters they capture the energy from filtered sunlight, convert it to carbohydrates, and serve as a food base for the near-shore ecosystem.

Eel-grass and surf-grass (*Phyllospadix* spp.) are vitally important to fish and other animals which have major economic value. Surf-grass occupies rocky shores on open California coasts. At exceptionally low tides it is exposed and reachable from land. One can then feel the tough, grass-like blades and see the sprawling rhizome system that clings tenaciously to rock surfaces and crevices.

Where tidal lands are shallow enough to lie exposed to mid-day heat for even short periods of time, eel-grass cannot survive. Here, a mudflat zone will exist between eel-grass beds and the lowest reaches of marsh, an area ecologically unsuited to both aquatic and terrestrial plants. The dominant plants are microscopic algae, unicellular bits of green life that attach themselves to the surface of the mud—mainly diatoms. The algae are abundant enough to form a soft, loose mat of tissue half an inch thick on the surface of the mud. Filaments of slimy blue-green algae, rough clumps of the green alga *Enteromorpha*, and delicate sheets of sea lettuce (*Ulva*) can also be present. The blue-green algae fix nitrogen, turning atmospheric nitrogen into organic forms, in addition to fixing carbon by photosynthesis. The organic nitrogen then becomes available to all life, passing through the food chains of both terrestrial marshes and aquatic harbors.

Coastal salt marshes form at the mouths of rivers, along bays, and in shallow depressions that periodically flood. Fine-grained sediments accumulate with the rise and fall of tides. Core samples from these clayrich deposits have shown that one to five inches of new sediment can be built up over a ten-year period in quiet estuaries and pools. Colonizing plants contribute large quantities of dead leaves, stems, and roots. Decomposition of this material creates a rich organic soil, black in color and impregnated with marsh gas (hydrogen sulfide).

Coastal Salt Marshes

California's tides are called "mixed, semi-diurnal" and have only a moderate range in elevation between highs and lows. Within a period slightly longer than twenty-four hours, there are two high tides (a high high and a low high) and two low tides (a low low and a high low). The average change in sea level between high high and low low is less than six feet. In other parts of the world, tides can be "simple"—one high and one low each day—and sea level can change as little as three feet or more than thirty feet.

Tidal marshes occupy gradually sloping ground, with elevation increasing inland. This means that the seaward, lowest portion of the marsh is inundated by high tides more frequently and for longer periods than the inland, upper portion. The plant community of the lower marsh is different from that of the upper marsh because wetland species differ in flood tolerance and salt tolerance. Soil salinity tends to be constant and moderately high (two to three percent) in low marsh because the soil is frequently bathed by sea water. The upper marsh episodically has a higher soil salinity: during long periods of exposure to the air between high tides, evaporating water leaves salt behind in the soil. When the upper marsh is finally washed by a high tide, salinity drops somewhat, and it drops even lower when leached by winter rains. Beyond the high marsh—that is, beyond the reach of highest tides—soil salinity declines to non-saline levels because rain and upland runoff dilute wind-blown salts that reach this far inland.

Close to the open water, where tidal inundation is most frequent and prolonged, the algal mat is the only vegetation. At a point where the hours of continuous submergence decline to nine and the hours of continuous exposure rise to thirty, cordgrass (*Spartina foliosa* and introduced *S. densiflora*) becomes established, and it dominates the lower edge or zone of the marsh from mean sea level to mean high tide. Cordgrass is a large native perennial that forms erect stands of bright green leaves and stems during the summer growing season. Cordgrass can become established by seed, but it generally expands by vegetative growth from rhizomes. The canopy emerges three or more feet above the soil and provides forty to sixty percent cover for a wide variety of waterfowl and even large aquatic animals such as sea lions. Nutrients released from decay of old leaves are a major source of food for offshore bacteria and small animals. These decay products (detritus) form the base of bay and harbor food webs.

The extensive root and rhizome system of cordgrass is almost constantly submerged. However, studies have shown that the hollow leaves, stems, and rhizomes are effective at passing oxygen from the atmosphere into the soil by simple diffusion. This specialized anatomy provides tolerance to flooding but not to salt.

Compared to other salt marsh plants, cordgrass is surprisingly sensitive to the salinity of tidal waters and soils. Its seeds germinate only when the low marsh has been sufficiently flooded with fresh water from rivers or rainfall. Since photosynthesis and growth are inhibited by salt concentrations that exceed the 3.4 percent of sea water, cordgrass grows only where tidal flushing keeps the root zone bathed in sea water. Where natural or human processes close a marsh to tidal action, it disappears. When cordgrass is absent, the low marsh is an extension of the algae-covered mud flat.

Cordgrass is not found all along the California coast. It is absent between Morro Bay and San Francisco Bay, and from Bodega Head north to Oregon except for exotic populations in Humboldt Bay of *Spartina densiflora* that arrived in the ballast of lumber ships from Chile a century ago.

An increase of only three to five inches in elevation is enough to reduce regular inundation and to promote conditions for the build-up of salt. With this increase an abrupt change in vegetation occurs: cordgrass of the low marsh is replaced by a more diverse mid-marsh community. Mid-marsh vegetation is a thick carpet of short, dense, succulent plants, and natural marshes are usually embossed with a pattern of meandering tidal channels. Here we find mainly perennial species that are extremely tolerant of root zone salinity. The most widespread and abundant of these is perennial pickleweed, which comprises sixty to eighty percent of mid-marsh plant cover. Its succulent stems have water-filled cells that isolate salt and keep it from interfering with sensitive physiological processes. Perennial pickleweed is more sensitive to waterlogging than low-marsh cordgrass, but in southern California annual pickleweed (*Salicornia bigelovii*) often grows with cordgrass. Other common mid-

Succulent stems of perennial pickleweed, a salt-accumulating species.

marsh plants include jaumea (*Jaumea carnosa*), arrowgrass (*Triglochin concinnum*, *T. maritima*), and saltwort (*Batis maritima*).

The upper marsh, only three feet higher than the lower marsh, is a third zone. Here continuous hours of submergence fall below five, and continous hours of exposure rise above fifty. The high marsh is irregularly flooded, and its soils show the greatest range of salinity, from nearly fresh during winter rains to twice the salinity of sea water during long periods of summer exposure. High-marsh species with succulent leaves include alkali heath (*Frankenia grandifolia*), tall pickleweed (*Salicornia subterminalis*), sea lavender or western marsh-rosemary (*Limonium californcum*), gumplant (*Grindelia stricta*), and sea blite (*Suaeda californica*).

Two perennial C_4 grasses native to the high marsh use specialized excretory glands instead of tissue succulence to regulate internal salt content. Saltgrass, with erect stems and stiff, angular leaves, is almost ubiquitous along tidal marshes of coastal California. Shoregrass (*Monanthochloe littoralis*), confined to southern California, has a low, creeping growth form that produces dense, spiny mats. Using large amounts of metabolic energy, the excretory glands of these

Saltgrass, a C_4 plant that excretes salt through specialized glands.

species collect and concentrate salts, releasing them as brine to the outside. White encrustations along the shoot indicate the position of the glands as the brine evaporates.

Annual plants are rare in the high marsh because they usually lack adaptations for tolerating salinity. Salt dodder (*Cuscuta salina*) and salt marsh bird's beak (*Cordylanthus maritimus* ssp. *maritimus*) are two exceptional annuals that parasitize perennial marsh plants. The mature pale orange stems of salt dodder lack chlorophyll and form dense tangles over host plants; they are divorced from the soil, having lost their roots after becoming established. Salt dodder develops minute knobs that invade host plants and form a connection to their water- and food-conducting tissues. Germination of salt marsh bird's beak is stimulated by rain, and the fresh water enables its roots to grow through the soil and connect with a host plant such as saltgrass. Bird's beak has salt-excreting glands and glistens with droplets of brine in the early morning.

It is easy to understand why so many species are excluded from this flooded, salt-dominated habitat. The structural and physiological adaptations for stress tolerance are metabolically expensive. But why do salt marsh plants not invade the upland where soils are aerated and non-saline? Experiments with the prostrate jaumea have shown that salt-tolerant plants do not compete well with upland species when there is little salt in the soil. In an experiment, seedlings of jaumea did not long survive when transplanted into grassland soil where tall, fast-growing grasses produced a dense, closed canopy. However, jaumea seedlings survived and grew well when upland grasses were hand-weeded from the plots. The adaptations that enable jaumea to thrive in the high marsh evidently do not allow it to grow rapidly. Upland species invest their resources to produce fast growth instead of tolerance to environmental stress. Competition, then, maintains the sharp boundaries between adjacent upland and lowland communities.

Brackish and Freshwater Marshes

Brackish and freshwater marshes are among the most productive ecosystems on earth, along with tropical rain forests and the most bountiful of temperate agricultural lands. The sheer abundance and high quality of plant food attract many different animals, both rare and common, as part of an intricate marsh food web. Humans are part of that web: we harvest large quantities of shellfish, fish, and waterfowl that feed on organic materials made abundant by brackish and freshwater marsh plants.

These marshes develop where freshwater runoff is sufficient to reduce tidal influence and prevent salt accumulation in the soil. Brackish marsh soils are moderately saline for some portion of the year. Freshwater marsh soils are low in salt year-round. Salt-sensitive species grow rapidly and reproduce in abundance when soils are least saline. Waterlogging is still a problem and specialized adaptations for avoiding root suffocation are common: hollow leaves, stems, and rhizomes to transport oxygen to roots and even biochemical innovations that reduce root consumption of oxygen. However, high rates of photosynthesis are sustained throughout the less saline spring-to-fall growing season and plants can grow to large size.

Brackish marshes do not have the low and high zones that characterize salt marshes. Local variation in species composition depends on differences in exposure to small amounts of salt during the summer growing season. Slightly saline areas are usually dominated by great bulrush (*Scirpus acutus*). Alkali bulrush (*Scirpus robustus*) and Olney bulrush (*Scirpus olneyi*) are less abundant and less widespread. Decreasing salinity permits common cattail (*Typha latifolia*), Baltic rush (*Juncus balticus*), and common reed (*Phragmites communis*) to mix with the bulrushes. All of these have the same erect, verdant aspect, and it is not easy to tell them apart from a distance; they are collectively called tules (*Scirpus acutus, S. californicus, S. validus*), and the tall vegetation dominated by these plants is commonly called tule marsh.

The brackish phase of tule marsh once covered more than 900 square miles around the Sacramento-San Joaquin River Delta of central California, forming a vast wetland landscape unmatched along the Pacific coast. Traveling through the delta in 1811, Padre Abella wrote in his journal:

> There are various delta islands covered with tule rushes and thickets. At fourteen leagues the rivers began to form, with tule on the banks. It is sheer swamp, which prevents any landing on firm ground. Everything is tule swamp on each side. . . the banks are covered with nothing but tule, and so high that one sees nothing but sky, water, and tule.

Only patches of this original delta landscape, totaling thirty-six square miles, remain. Smaller, less representative examples of brackish marsh may also be seen in south San Diego Bay and within major estuaries of north coast rivers, where an abundance of fresh water flows to the ocean. Even these small remnants, however, are endangered by water diversion, filling, and nearby development.

Great bulrush is somewhat tolerant of low salinity and very tolerant of flooding. It begins growth from deep rhizomes in late spring and produces stout

A mix of brackish and freshwater marsh species, including great bulrush and common cattail in the center.

Fruiting stalk of common cattail, a dominant perennial plant of freshwater marshes.

stems more than six feet tall by mid-summer. The dense, dark green canopy of great bulrush provides more than forty percent cover along river edges and estuaries. Its seeds ripen by autumn and are the single most important food resource for millions of migrating waterfowl.

Vegetative growth and seed production in great bulrush depend heavily on the availability of fresh water to prevent salt build-up in the soil. Lack of fresh water heightens the effects of long-term climatic shifts that reduce rainfall and river flow and the diversion of upstream water for human use.

During the drought of 1976-77 salt water intruded into the tule marshes of the Sacramento-San Joaquin Delta. Ecologists monitored the marsh plants and compared their distribution, abundance, and growth during drought and nondrought years. Great bulrush decreased in abundance and its stems grew less than half as much as they did in the previous non-drought year. Some areas had no bulrush growth at all, allowing pickleweed and other salt marsh species to spread into the once brackish marsh. Other tule species were even more seriously affected, creating a dramatic contraction of brackish marsh in just one year. Recovery from the drought was not difficult because these rhizomatous species could sprout, using stored food, once the salts were leached away by spring flooding. Recovery from continuous and progressive reductions in freshwater flow, however, is much less certain. Rhizomes depleted of their food reserves will eventually die after several years of high salinity.

Freshwater marsh plants are restricted by variations in water depth and swiftness of the current, rather than by fluctuations or gradients in salinity. Near flowing open water the depth of the water may prevent rooting of even the tallest species; here only floating, aquatic plants such as native duckweed (*Lemna* spp.) and introduced water hyacinth (*Eichhornia crassipes*) display themselves. If the current is sluggish, the dense cover creates the appearance of a river choked by its own plant life. As water depth diminishes toward shore, common cattail forms a thicket of strap-like leaf blades and "corndog" floral stalks. Tolerant of flooding but sensitive to salt, cattail dominates freshwater marsh but is only a minor component of brackish marsh. Great bulrush often accompanies cattail, along with rushes, sedges, bur-reed (*Sparganium eurycarpum*), and common reed.

In shallow areas, a low canopy of herbs and shrubs forms patches among the freshwater tules. These species have broad leaves arranged along prostrate stems: marsh pennywort (*Hydrocotyle* spp.), Pacific silverweed (*Potentilla egedii*), swamp knotweed (*Polygonum coccineum*), and miner's monkeyflower (*Mimulus guttatus*).

In northern California the edges of freshwater marsh typically blend with species from riparian forest (Chapter 4). Freshwater marshes are rare in southern California due to low rainfall and a lack of year-round streams.

The Changing Landscape

Prior to 1900 four to five million acres of wetlands existed in the state of California. Today only 450,000 acres, about ten percent of the original pristine area, remain and much of this acreage is degraded. Correlated with this reduction in wetlands has been the decline of shellfish, fish, and waterfowl resources that were once found in such abundance. Water development by the U.S. Bureau of Reclamation and the State Water Project reduced the geographic extent of flooding with dams, levees, drainage channels, and sophisticated water transport systems. These technological feats allowed land to be cleared, cultivated, and populated. Islands were absorbed into the mainland, entire lakes disappeared, and rivers no longer flowed in their lowest reaches. Marsh vegetation gave way to salt ponds, orchards, and row crops, which then gave way to houses, roads, and shopping malls.

The wetlands that remain are in poor condition. Sixty-two percent of existing coastal wetlands are rated by the California Coastal Commission as severely damaged. Nearly sixty percent of the water that once flowed into the wetlands of the Sacramento-San Joaquin River Delta is now stored and diverted for agricultural, residential, and industrial purposes. If all projected demands for delta water are met, another ten percent will be diverted by the year 2000.

Diversion allows tidal waters to penetrate farther upstream, bringing additional salt to areas of brackish and freshwater marsh. Progressive salinization alters the composition and integrity of these kinds of wetland vegetation by pushing species beyond their tolerance limits for salt- and drought-stress. Complete flushing of the wetlands is already infrequent, resulting in a detrimental accumulation of salts. Additional water diversions will have serious effects on the remaining wetland vegetation of the Sacramento-San Joaquin River Delta.

Freshwater flow from rivers to the delta is also important for the maintenance of salmon populations. During the early 1990s, after five consecutive years of drought, plans were made to release water from dams along the Sacramento River so that migrating salmon could survive. Of course, water use for ecosystem conservation represents a loss of water for agriculture and recreation. We must reconsider our priorities for the allocation of water. Currently, the demand of agricultural and urban systems for water is ranked ahead of that for natural ecosystems. In the face of a finite and diminishing supply of water, this choice will lead to environmental degradation.

Ironically, as wetlands disappear from California and the nation we are realizing their great biological, economic, and cultural values. Wetland vegetation dissipates floodwaters, absorbs large quantities of nutrients that pollute surface water, and helps to recharge groundwater supplies. Wetlands provide food and nesting sites for millions of waterfowl and shorebirds along the Pacific flyway. They form the basis of the food web that supports many commercially important species of fish and shellfish. They represent recreational lands for fishermen, duck hunters, river lovers, and naturalists. They har-

The restored Muzzi Marsh (Marin Co.) now has extensive stands of cordgrass and is home to a small population of the endangered clapper rail. Photograph 1990.

bor endangered species of plants and animals, most of which are uniquely adapted to saline or flooded conditions. Salt marsh bird's beak, Suisun thistle (*Cirsium hydrophilum* var. *hydrophilum*), and delta tule pea (*Lathrys jepsonii* ssp. *jepsonii*) are only a few of the wetland plant species threatened by extinction. Endangered animals such as the salt marsh harvest mouse and California clapper rail also depend on the few remaining scraps of marsh. Twenty-five percent of the plants and fifty-five percent of the animals designated by the state as either threatened or endangered have wetlands as their essential habitat. Furthermore, wetlands are places of great openness and perspective, with sweeping panoramas that offer a simple beauty which encourages human reflection.

Awareness of these threats and values has inspired efforts by private and public organizations to restore wetland vegetation. Across the nation pioneering projects have established the importance of returning to natural cycles of inundation, salinity, and sedimentation in order to create suitable habitat for wetland plants. An early restoration project was carried out by the Golden Gate Bridge, Highway, and Transportation District as mitigation for dredging a channel to a new ferry terminal in San Francisco Bay. A barren 200-acre site that had been diked from tidal activity in the 1950s was acquired in 1975—in part to accept dredge spoils and in part as a mitigation site. The dikes were breached in 1976, restoring tidal circulation to 130 acres of the mitigation site. Within three years, new vegetation covered much of the site, called the Muzzi Marsh. Additional channels were dug in 1980 to reduce access to the marsh by dogs, cats, and motorcycle vandals and to increase the reach and volume of tidal water to landward portions of the marsh.

Long-term monitoring of the Muzzi Marsh has clarified some general lessons about marsh restoration. We have learned, for example, that plants establish themselves readily when soil conditions are favorable, without expensive planting programs; and that minor changes in topography and elevation are critical to successful revegetation. Today, restored portions of the Muzzi Marsh look like a pristine wetland. Ongoing monitoring is quantifying how much of the total ecosystem function has returned, compared to an undisturbed ancient tidal marsh. Restoration certainly seems to offer some hope for increasing this type of endangered habitat.

There are, however, at least three major obstacles that stand in the way of reclaiming large tracts of wetlands. First, there are few remaining lowland sites adjacent to water at elevations suitable for restoration because most low-elevation areas have been developed. Second, most of the remaining land that would be suitable for wetland restoration is so commercially valuable that government agencies and conservation groups can seldom afford to buy it. And finally, restoration itself can be expensive, costing several thousand dollars per acre if significant grading is involved. At the present time, restoration technology at least should be used to repair and expand existing natural wetlands that have been preserved as refuges, parks, and watersheds. Some local wetlands, of course, can be inexpensively reclaimed by merely opening dikes; usually, however, the duration and extent of degradation are too severe to be so simply reversed.

Although preservation of wetlands has finally become a national conservation issue, it has not been effectively addressed by legislation. The Emergency Wetlands Resources Act, signed into law by President Reagan in November 1987, requires states to enlarge their land use planning process to include wetlands, but it proposes no direct action to reduce the rate of habitat degradation. Unfortunately, most of the wetlands remaining in California are privately owned and may not be protected under existing statutes. Even worse, there are government programs that effectively subsidize the destruction of wetlands on private land by promoting clearing, draining, and conversion for "useful" purposes. Federal and state governments, as well as private organizations such as The Nature Conservancy, are acquiring intact, healthy wetlands, but funds are limited and subject to great political struggle.

Parcels of protected marsh are not autonomous islands; they depend on water flows from elsewhere. Lateral flow of fresh water into salt and brackish marshes, for example, is critical to the maintenance of characteristic vegetation. Diversion of freshwater flows away from wetlands dramatically changes marsh vegetation and the animal life within it—invertebrates, fish, amphibians, wildfowl, and mammals. The struggle to provide that water for wetland conservation is just as great and just as political as the struggle to preserve the land. Those struggles can be won only with enthusiastic support from public and private sectors of a society that recognizes the great value of wetlands.

3

COASTAL FORESTS

THE COASTAL MOUNTAINS ARE A FOG FENCE, A BARRIER SEPARA-
ting the cool, foggy maritime strip from the hot, arid interior.
Within the pool of fog, coddled by the moderate environment, lies a
unique forest that is a living fossil from ancient climates: the coast
redwood forest. It is a rich mix of species that have come together over the course
of tens of millions of years. This forest from another time forms the southern tip

of a magnificent conifer forest that blankets
a strip of coastal slopes and flats from Alaska
to Monterey County. The humid, cool,
maritime climate bathes the trees and creates the fastest growing forests in the world.

Redwood trees (*Sequoia sempervirens*)
loom above the forest floor, enveloped in
fog and stilling the air beneath gray-green
canopies. Three hundred feet below their
branches, myriad ferns and forbs and low
shrubs cluster among the massive buttresses,
forming a soft, silent, dripping wet under-
story. The forest is easy to move through on foot: shrubs of huckleberry (*Vaccinium
ovatum*, *V. parvifolium*) and salal (*Gaultheria shallon*) are scattered and yielding;
understory trees of wax myrtle (*Myrica californica*) send slender, twisting stems
upward; herbs of sword fern (*Polystichum munitum*), redwood sorrel (*Oxalis
oregana*), maianthemum (*Maianthemum dilatatum*), and false Solomon's seal
(*Smilacina racemosa*) form a mosaic of carpets and openings.

The forest floor is dim, and color is uncommon; so the pastel pink of redwood
orchid (*Calypso bulbosa*) flowers, the deep red of clintonia (*Clintonia andrewsiana*)
blossoms, and the bright white undersides of trail plant (*Adenocaulon bicolor*)
leaves are startling. Members of this forest have close relatives in the Sierra
Nevada, the eastern United States, northern Europe, China, Japan, and British
Columbia. Their ancestors were part of the Arcto-Tertiary geoflora and of a great
northern-temperate forest that once encircled the globe.

*Redwood forest (Jedediah
Smith State Park, Del
Norte Co.) (opposite), with
massive redwood trees in
the overstory, lesser
conifers and tall shrubs in
the understories, and an
herbaceous layer of sword
ferns and redwood sorrel
near the ground.*

*Redwood sorrel (above),
illuminated by a fleck of
sun on the forest floor.*

Though the forest is seemingly placid, violent and noisy episodes of fire and flood sweep through it every century. Young trees, shrubs, and herbs are periodically consumed by fire or buried in layers of fresh silt. The mature redwoods survive. Thanks to a unique root system and an insulating layer of bark, young trees germinate on the fresh soil, and the forest continues its unbroken existence. Westerlies from the ocean bring banks of summer fog through the forest, fog that turns to nourishing rain as the mist condenses on branches and falls as droplets to the ground. There is no hot, dry summer in the redwood forest.

Green Gold: Coast Redwood

The first record of people other than Native Americans seeing redwood forest comes from the Portola expedition in Santa Cruz County in 1769. Only fifty years later Russians at Fort Ross were harvesting redwood on a small scale and exporting lumber to Hawaii. The economic value of these forests and the quality of redwood lumber became apparent to Europeans when gold rush immigrants soured on mining. Some harvested timber—green gold. Pine (*Pinus* spp.) and fir (*Abies* spp.) were cut in the Sierra Nevada for mines and homes, but redwood was the premier lumber tree in the Coast Ranges. Here was a tree with a tall, straight bole and few knots, growing in open and sometimes level forests, within easy distance of shipping ports.

In 1850 two million acres of mature redwood forests formed a strip five to thirty-five miles wide and 100 to 2,000 feet in elevation along the coast from the Oregon border to the southern boundary of Monterey County. This represents two percent of California's land mass. It was a quiet, still, dripping-damp, dimly lit forest of heroic proportions. The redwood trees are among the tallest in the world, reaching 369 feet. Growth is also exceptional. At maturity (400-500 years old), these trees become more than 200 feet tall and twelve feet across at the base. Old-growth stands contain hundreds of thousands of board feet per acre, ten to 100 times the volume of any other California forest. The wood is excellent for construction because it is impregnated with natural compounds such as tannins that retard fungal and bacterial decay. These compounds also contribute to the longevity of the uncut trees, which reach maximum ages of 2,200 years.

Exploitation of this resource grew slowly because handling timber the size of redwood logs required steam-powered sawmills, shipping ports, and interior roads or waterways to transport logs to the mills. It also required fleets of coasting ships to take finished lumber to markets in San Francisco. All of these requirements were met along the north coast only in the 1850s, and even then some were barely met. The rocky north coast offered poor protection from the open ocean, and considerable sailing skill was necessary to load ships along the coast. Seamen disparagingly called north coast timber stops "dog hole" ports.

Logging was criminally wasteful in the nineteenth century. The swollen bases of trees were too large to cut through, so fallers drove springboard planks into the trunks ten feet or more off the ground, and stood on them to chop and saw even higher. Trunks were so enormous that it took two men, standing on opposite planks, several days to fell a tree. Once on the ground, buckers removed bark and branches and sawed trunks into ten to twelve foot lengths. Logs were dragged by teams of oxen along skid roads constructed of small-diameter logs. If the distance to a mill was longer than a couple of miles, waterways were modified to be used for transportation instead of skid roads. Logs were then rolled into creeks. Small dams above them were released in winter, to carry the logs on crashing flood waters down toward the ocean. Erosion was severe. Tree trunks, slash, and uncut timber were burned after logging took out the best trees, and these uncontrolled fires caused additional erosion.

Long teams of strong oxen were needed to move redwood logs out of the forest and on to the mills of Mendocino County in 1870.

But wood was plentiful and the demand in San Francisco seemed endless. Wood was unrealistically cheap, and no thought was given to the ecological cost. Redwood facing was so common in this period that homeowners painted it to look like stone. The timber industry was sustained by San Francisco's habit of burning to the ground every ten years during the 1800s. After 1880 steam-driven Dohlbeer donkey engines replaced oxen for hauling. The new technology increased erosion because logs could be hauled over steeper terrain. Further mechanization in the twentieth century accelerated the pace of harvest, and only ten percent of the pristine redwood forest that existed in 1850 remains today. By the year 2000 five percent or less of that original acreage will exist.

A 500-Year Management Plan for Redwood Forest

Humans have the wrong life span to be natural conservationists: it is either too short or too long. If we lived only a decade, we could never develop the ability or vision to organize development projects on the scale we do now. Dam construction, large cities, highways, and complex agricultural systems would not exist because they need more time than a ten-year lifespan provides. If we lived 200 years we would have a long-term view of resources. But instead of ten years or 200 years, we live an average of seventy: just long enough to do considerable ecological damage, and short enough to lack a long-range conservation outlook. Of course, we can always learn to be conservationists, but that takes altruism and altered behavior that doesn't come naturally.

Conservation has had several meanings. One definition equates it with preservation. That is, natural resources are to be locked up, left alone, protected from exploitation and even from such natural disturbance as fire. For some ecosystems, this version of conservation is suitable, but redwood forest cannot be isolated from disturbance. Its reproduction and continued successful growth depends upon periodic fire and flood—not devastating crown fires, but less intense, cooler burning ground fires. Redwood trees have structures which protect them from this sort of fire.

Fire-resistant, insulating redwood bark prevents serious injury to most mature trees. Surviving root crowns at the base of the tree contain buds buried beneath

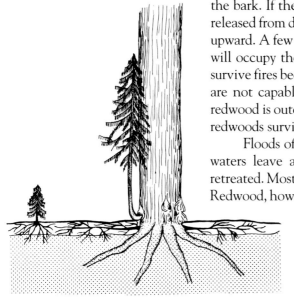

the bark. If the top of the tree has been killed by fire, these buds are released from dormancy. A cylinder of sucker shoots emerge and grow upward. A few of these will mature into a circle of daughter trees that will occupy the space of the dead parent. Other tree species do not survive fires because their bark is more flammable or their root crowns are not capable of sending up sucker shoots. If no fires occur, the redwood is outcompeted by other tree species. When fire occurs, only redwoods survive.

Floods often follow fires, because the land is more open. Flood waters leave a thick deposit of fresh silt behind after they have retreated. Most plant roots are suffocated by flood water and sediment. Redwood, however, tolerates the inundation and sends new roots into the layer of silt that grow through the bark along the buried trunk. No other tree species has this capability. Leaf litter is also buried by the silt. Litter prevents young redwood seedlings from surviving, so burial of the litter by flooding and its consumption by fire increase seedling establishment.

Coast redwood is capable of tolerating flood and fire in several ways. New roots can form in fresh silt deposited by swollen rivers. Sucker shoots can sprout from buds in the base of the parent's trunk if the above-ground portion has been killed by fire. Seedlings, left, also establish readily in fresh silt or after a ground fire.

Historical evidence of fire and flood is found in the soil beneath redwood forests. Excavation reveals successive layers of carbonized wood and silt, each layer of charred wood separated from the next by a sandwich of silt. Evidently, fires or floods have occurred every thirty to sixty years for at least the past millenium. However, in the last century, fire suppression and flood control have virtually eliminated these natural disturbances.

In the absence of fire or flood, saplings of other species will grow up beneath the redwood canopy. As future centuries pass, redwood trees will reach their natural span of years and die from disease or toppling in high winds. When they fall, a hole in the canopy rich in light and growing room will be created. Any saplings in that gap will grow rapidly and eventually reach the canopy, filling the gap. Species that thrive in these openings include Douglas-fir (*Pseudotsuga menziesii*), tanbark oak (*Lithocarpus densiflorus*), Sitka spruce (*Picea sitchensis*), grand fir (*Abies grandis*), and cedar (*Chamaecyparis* and *Thuja* spp.). If we continue to suppress fire and flood, every old-growth redwood tree will eventually fall and the community will become extinct. The new climax will be a fir-oak-cedar forest.

Ironically, if preservationists do everything they can to protect redwoods from fire and flood, they will eventually exterminate these trees. Park managers must permit, encourage—even create—a suitable ground fire or silt-laying flood often enough to maintain the dominance of redwood. Although the frequency of natural disturbances was thirty to sixty years, a managed frequency could be as seldom as 100 to 500 years. A 500-year management plan is quite a challenge for humans used to thinking in terms of next weekend or next year, but old-growth redwood forests will not persist unless humans meet that challenge.

Most commercial logging is clear cutting: that is, loggers enter a site once, remove all marketable timber, and burn the slash. The shape and area harvested may be chosen to maximize the rain of seeds from adjacent forest, or to provide shade or wind shelter for the bare area. Care is taken to minimize erosion. Nevertheless, clear cutting alters the microenvironment enormously. Light intensity on the ground increases ten-fold, causing daily temperature extremes there to skyrocket and soil moisture to decline. Species that once existed in the overstory may not grow as well as seedlings in the clear cut.

Clear Cut or Select Cut?

Redwoods require an almost constant temperature for best growth. Experiments with seedlings show that 63°F day and night is optimal. Redwoods also require moist soil. Seedlings beneath a natural cover of parental trees are shaded and grow in a constantly cool, humid microclimate. Summer fog passing through the canopy condenses on the needles and drips down to the ground. Fog drip during one summer is equivalent to eight inches or more of rainfall. Also, fog drip is richer in nutrients than rain water, because it accumulates nutrients as it runs along leaf or bark surfaces. Seedlings in clear cuts do not experience moderate temperatures, and they do not receive summer supplements of moisture and nutrients from fog drip.

Because redwood seedlings do not thrive in open clear cuts, Douglas-fir seedlings are being planted instead by land managers. Redwood stumps are not usually removed, however, so stump sprouts mix with Douglas-fir saplings in the new forest. Douglas-fir, redwood stump sprouts, and some redwood seedlings grow to merchantable size within seventy years.

An alternative timber harvesting technique is select cut, where some fraction of the taller trees are marked and removed. Ten or so years later the site is reentered and another fraction of overstory trees removed, and so on. After seventy years, the same volume of timber has been removed as in the clear cut method, but the site has never been laid bare and exposed to the macroenvironment. Select cut should lessen erosion and increase natural regeneration, although repeated entry disturbs the site almost continuously and some damage to unharvested trees and saplings and seedlings inevitably occurs. Trees do not always fall where aimed by loggers, and they can damage other trees on their way down. Heavy equipment is large and awkward, and it too can injure trees. Sometimes the scattered trees remaining are more susceptible to windthrow than a closed forest, and additional trees are lost.

For these reasons foresters are coming to favor clear cut techniques for redwood, modified for minimal erosion and maximum recovery. Many foresters think second growth could be clear cut again and the cycle repeated. Some redwood forests have been harvested twice, but we have no idea how many times the cycle can be repeated. Contrary to the widely held notion that forests are renewable, we don't have enough experience yet to know whether short-term rotations can be sustained without long-term ecological degradation and expensive site improvements.

As just one example of improvements that may be needed, it is possible that nutrients such as nitrogen will have to be added to maintain a seventy-year cycle. Fertilization is an expensive procedure and the cost of wood would rise. Runoff waters would be enriched in nutrients, and pollution problems downstream are sure to result. At present, we are treating forest timber like ore in a mine, but in the future we may have to treat it like corn in a field or, finally, like a rare animal in a zoo.

Mixed Evergreen Forest

The mixed evergreen forest is a sandwich layer between slices of low and high elevation forests. It grows on steeper, warmer, and higher slopes, ascends the west flank of the Coast Ranges, spills over 5,000-foot elevation ridges, and runs down interior-facing slopes eastward to the edges of foothill oak woodland and chaparral. Less majestic than redwood forest, but twice as extensive in acreage, this mixed evergreen forest covers five percent of California's area.

In the wettest part of California—the Klamath Mountains of the northwest—mixed evergreen forest occurs between 1,000 and 4,000 feet elevation. Above 4,000 feet montane conifer forest replaces mixed evergreen forest. The evergreen forest is intolerant of serpentine soil, so serpentine slopes within that elevation belt are dominated by brush fields with scattered conifers. South of Monterey County the coastal strip does not contain redwoods. Here mixed evergreen forest lies just above coast oak woodland and chaparral at elevations between 2,000 and 5,000 feet. In the Transverse and Peninsular ranges of southern California, mixed evergreen forest lies between 3,500 and 5,700 feet elevation. The mild climate of this mid-elevation zone in California is characterized by less annual rainfall than the redwood forest, and a greater range of daily and seasonal temperatures.

Like any sandwich, the mixed evergreen forest contains flavors of the slices above and below it. It contains plants found at low and high elevations in addition to species unique to its own zone. Dominant trees include needle-leaf evergreens (conifers) and broadleaved evergreens. Needle-leaf tree species change with latitude. Typically they are Douglas-fir in the North Coast Range and Coulter pine (*Pinus coulteri*) from Monterey south. Broadleaved trees usually include coast live oak (*Quercus agrifolia*) or canyon oak (*Q. chrysolepis*), accompanied by other species with narrower ranges.

Northern Mixed Evergreen Forest: Douglas-Fir Hardwoods

Douglas-fir dominates the lower montane zone of California's North Coast ranges. Along with redwood, it belongs to the Pacific Northwest biotic province, but its range is more extensive than the redwood, spilling east into the Cascade Range and the Rocky Mountains. Its best growth occurs near the coast, large individuals there reaching over fourteen feet in trunk diameter, over 250 feet in height, and more than 1,000 years in age. Typical mature trees are five to seven feet in diameter, 200 feet tall, and 600 years old.

In the cool and wet Pacific Northwest, Cascades, and Rockies, Douglas-fir forms a closed overstory. In drier and warmer California, however, it forms a ragged, patchy, incomplete canopy that allows full sun to reach down between the trees. Sunlight is intercepted by a continuous cover of broadleaved trees, far below the conifer tops, standing only forty to sixty feet tall: tanbark oak, madrone

(*Arbutus menziesii*), bigleaf maple (*Acer macrophyllum*), California bay (*Umbellularia californica*), California black oak (*Quercus kelloggii*), coast live oak, interior live oak (*Quercus wislizenii* var. *wislizenii*), canyon oak (*Quercus chrysolepis*), hazelnut (*Corylus cornuta* var. *californica*), and mountain dogwood (*Cornus nuttallii*).

The richness of this vegetation type is mostly in its trees, because shrubs, mosses, and perennial herbs are relatively few. A thick bed of slowly decomposing leaf litter completely covers the ground. Most of the tree cover is evergreen, but maple, black oak, hazelnut, and dogwood are winter-deciduous and their leaves contribute golden, yellow, and purple hues to the fall landscape. Proximity to the ocean and moderate elevations confer gentle wintertime conditions favorable to evergreens. Evergreens are able to grow actively twelve months of the year. Deciduous trees are dormant for several months, so they are at a competitive disadvantage. Leaflessness in winter would be an advantage only in the face of

Away from the wet coastal mountains, mixed evergreen forest is a low, dense assemblage of oaks, madrone, and bay that often borders on oak woodland and grassland.

hard frost or heavy snow. Snowfalls do occasionally occur, and they damage the broad-leaved evergreens, whose leafy limbs intercept too much snow and break off under the weight.

Douglas fir is the most commonly cut timber tree in North America. It once was a major element in the mixed evergreen forest of northern California. Because of logging, old growth Douglas fir forests are now rare. The Nature Conservancy's 5,000 acre North Coast Preserve in Mendocino County is the largest single old-growth tract left in the state—probably containing half of the entire state's old-growth acreage. Timber harvesting in mixed evergreen forest is currently both an economic and an ecological disaster, because it can't be accomplished on a sustained, repeatable cycle. Understory hardwoods—especially tanbark oak and madrone—are capable of stump sprouting following disturbance. They regenerate enthusiastically after logging. Many stems arise close together and competition among them results in a "dog hair" stand composed of as many as 1,000 spindly trees per acre.

Douglas fir cannot stump sprout and must come back from logging by seed. Young seedlings and saplings are close enough to the ground to be grazed by animals. Since deer and domestic livestock prefer Douglas fir to hardwood foliage, grazing intensity can be so high that conifer seedlings will not survive unless protected by wire fencing. Fencing is an unrealistic expense to adopt on the large scale of a commercial logging operation.

The thousand stems of hardwood stump sprouts per acre put a severe competitive stress on Douglas fir saplings. Tanbark oak and madrone are nearly worthless as timber species in today's market. Often, herbicides are applied to kill hardwood species, releasing Douglas fir from competition. They are killed by injecting herbicide into their trunks. Small scale tests have shown that the growth of Douglas fir stems triples if the hardwood density can be reduced by eighty percent . This procedure, however, is too expensive to be employed on a large scale.

Old-growth forests are not ecologically the same as young second growth forests, even if they share all the same species. The architecture of the young forest is different from mature forest. The density of each canopy layer, the size of emergent trees, and the pattern of tree distribution are not what they were a century ago, and these differences are important to animals. Old-growth Douglas fir forests are the only habitat of some rare animals, such as spotted owls, marbled murelets, and red tree voles, which nest in or on large Douglas firs or dead snags. Old-growth forests also contain large amounts of standing or fallen dead wood that provides habitat for a diversity of insects and fungi. Certain species of salamanders are also found only on the wet floor of old-growth Douglas fir forests. The juvenile forest that follows logging or fire does not contain these organisms.

Some foresters disparagingly call old-growth forests "over-mature." The implication is that they are senescent, slow-growing shadows of their once grand

Northern mixed evergreen forest along the south fork of the Eel River (Mendocino Co.), dominated by Douglas fir and broadleaf evergreen trees. A few redwood trees are also present in this wet portion of the North Coast Range.

selves. It is true that old growth trees grow slower than young trees, but the species richness and complexity of old growth forests are immeasurably greater. Old-growth forests are banks of genetic resources, resources not found in second-growth, younger forests.

Southern Phase of Mixed Evergreen Forest: Coulter Pine-Hardwoods

Southern mixed evergreen forest on Mt. Palomar (San Diego Co.), dominated by big-cone spruce, Coulter pine, canyon oak, and California black oak.

South of Monterey County, Douglas fir is no longer an important member of the mixed evergreen forest. Some southern stands of mixed evergreen forest lack a conifer overstory and are composed completely of hardwoods, but most contain Coulter pine or big-cone spruce (*Pseudotsuga macrocarpa*). Associated with the conifers are canyon oak, coast live oak, California black oak, and a scattering of evergreen shrubs. Coulter pine has long, yellow-green needles in three's and has a growth form that mimics ponderosa pine. Its cone, however, is unique: the most massive cone of any conifer in California, about twelve inches long and eight inches thick, weighing over two pounds, and with wicked recurved spines at the end of each cone scale.

Ecologically, Coulter pine is one of the mystery trees of California. Very little information is known about its biology, despite its widespread importance to vegetation. It ranges south from Mount Diablo through the Coast Ranges, at first

in scattered locations, then over larger and larger continuous areas as it spills into the Transverse and Peninsular Ranges. Where it grows near woodlands, which burn much less frequently than the mixed evergreen forest, the cones open at maturity. Where its grows near vegetation that burns frequently, such as chaparral, its cones are closed and open only after fire.

These typically irregular canopies of big-cone spruce are emerging from a chaparral dominated by evergreen shrubs (Strawberry Peak, L.A. Co.).

Massive seed release after fire is a rare behavior for conifers. Most of the world's conifers have cones which open when mature, allowing seed dispersal by wind and gravity. But about one third of California's conifers—eighteen species of pines and cypresses—do not release their seeds when the cones mature. These are the closed-cone conifers. The cones remain sealed shut by a resinous glue and they also remain attached to the tree. If the cone is retained for decades because fire does not occur, the cones become half-buried in bark. While a few cones may be opened by animals, by decay, or after the passage of many years, only high temperatures such as those resulting from fire open the cones in mass. The resin melts and burns, and the scales begin to open, exposing seeds within. The cone insulates most of the seeds from high temperature, so they remain alive.

Most conifer species are either closed-cone or open-cone. Coulter pine is unusual in having some individuals with open cones, some with closed cones. Coulter pine grows on hot, arid, south-facing slopes surrounded by flammable brush, and fire frequently moves through this habitat. In contrast, the other common southern conifer of mixed evergreen forest is big-cone spruce, which has open cones and exists in relatively fire-free habitats such as ravines or cool, north-facing slopes.

Southern Oak Woodlands

Several woodland plant communities unique to southern California occur just below the mixed evergreen forest. We include them here, rather than in Chapter 4, because they are part of the fog fence. They grow on west-facing slopes within the maritime climate zone. Mean annual temperatures here are a few degrees warmer, and mean annual rainfall more than ten inches lower, than in the mixed evergreen forest above.

The woodlands are dominated by coast live oak, mesa (*Engelmann*) oak (*Quercus engelmannii*), and California black oak. They spread over coast-facing or north-facing slopes below 3,800 feet elevation. The architecture of the vegetation is typical of woodlands anywhere in California. Scattered trees thirty to seventy feet tall and less than a foot in trunk diameter shade a nearly continuous carpet of herbs. Some stands are savanna-like, with fewer than a dozen mature trees per acre, but typical woodland stands contain up to sixty trees per acre. In these stands, coast live oak is mixed with the deciduous trees California black oak and mesa oak. Oak-walnut woodlands are threatened with extinction because they have been disturbed, modified, or entirely displaced by cities and suburbs. The Nature Conservancy and the State of California have put southern woodlands among the state's thirty-two most endangered plant communities, in the highest priority class for inclusion in protected natural areas.

Fire Pines and Cypresses

Prior to the 20th century, wildfires were a natural part of the environment in three-fourths of the state's area, and they occurred with a regu-

Cones of Bishop pine open with age, hot weather, or fire.

lar frequency. Stands of closed-cone conifers burned once every twenty-five to fifty years. Some closed-cone conifers, such as Monterey pine (*Pinus radiata*), have cones which will open with age or on hot days when the cones are in full sun, but only fire

A native, fire-prone community of Monterey cypress and Monterey pine at Point Lobos State Reserve (Monterey Co.).

creates high enough temperatures to open the cones of most closed-cone species. We've already described Coulter pine as having some trees that open their cones normally, whereas other trees require the heat of a wildfire. Perhaps the most fire-dependent species is knobcone pine (*Pinus attenuata*), which grows throughout the coast ranges. It has cones which open only when exposed to temperatures near 200°F for about five minutes.

Groves of closed-cone conifers are usually dense and difficult to walk through because lower limbs stay attached to the trunk. These limbs add to the fuel load and act like a ladder for flames on the ground to climb upon. There may also be

a dense understory of flammable shrubs. The groves are typically islands within surrounding vegetation that burns often, so the probability of fire starting in a grove or in adjacent scrub is high. Any fire which reaches a closed-cone conifer grove quickly builds into an all-consuming crown fire. Only charred tree skeletons remain amid the soundless fall of an immense seed rain onto a deep bed of nutritious ash. The seeds will germinate the following wet season and begin the existence of a new grove. Any seeds of other plants arriving later will be at a disadvantage, because of the head start by conifers.

Most closed-cone conifers are coastal in their distribution, and most grow on droughty or nutritionally poor sites with sandy, salty, acidic, shallow, or serpentine soils. They are relatively small trees with life spans of 75-300 years. A few species are well known because they grow near cities; they have ornamental or economic value, or they have some peculiar growth form. For example, the Monterey Peninsula landscape owes its mystique to sweeping silhouettes of Monterey cypress (Cupressus macrocarpa) along a rocky coast. This species is now widely planted as an ornamental, but it has a very limited natural distribution, as does Monterey pine, both of which grow wild only along the windy, salty, foggy Monterey coast. In many areas outside of its natural coastal distribution where environmental stresses are high, Monterey pine is one of the fastest-growing tree species of the temperate zone. Extensive plantations of Monterey pine thrive in Argentina, Australia, Chile, Kenya, New Zealand, South Africa, Spain, and Uruguay.

The evolutionary routes to modern niches are surely intricate and varied, but one of the most bizarre end products must be that of the closed-cone conifer. Here is a plant whose offspring survive only if the parental generation is destroyed. The parents cling together in tight, fire-prone groves that positively increase the chance of their destruction. It's akin to premeditated suicide. They have evolved so intimate a dependence on destruction that their life cycle is incomplete without it.

Pygmy Plants on a Giant Staircase

The strangest looking closed-cone conifers are Bolander pine (*Pinus contorta* var. *bolanderi*) and pygmy cypress (*Cupressus pygmaea*). These natural bonsai trees, only two to four feet tall and covered with lichens, create a dwarf forest on the poorest, most acidic of soils in Mendocino and Sonoma Counties. For the past half million years, sea level has risen and fallen hundreds of feet as glaciers melted and grew several times. Each level of the sea built a shelf below the eroding, crashing surf. The shelves were heaved above-ground in a succession of uplifts that give the modern coast landscape a terraced appearance. Sea level change and the creation of terraces happened all along the California coast, but they have been best preserved in parts of Mendocino County.

The terraces resemble a giant staircase, each step upward and inland being about a half-mile wide, 100 feet above the next, and roughly 100,000 years older. The most landward steps—the fourth and fifth terraces—are three to five miles inland, 400-600 feet above sea level, and have Blacklock soils atop them nearly half a million years old. These may be the oldest soils in the world, and they are among the worst for plants to grow on. Half a million years worth of weathering have produced an acidic, nutrient-poor, shallow soil that results in every plant on it becoming dwarfed. A pygmy forest, with an open canopy of Bolander pine and pygmy cypress, pokes up among dwarf shrubs of salal, huckleberry, rhododendron (*Rhododendron macrophyllum*), labrador tea (*Ledum glandulosum*), pygmy manzanita (*Arctostaphylos nummularia*), and tufts of reindeer lichen.

Why is the soil so poor? Percolating rainwater has leached the upper soil horizons of clay, organic matter, most essential nutrients, and bases. Soil pH is 2.8-4.0, a thousand times more acidic than good agricultural soil. Part of the upper horizon is bleached white in color because it has been so thoroughly leached. Below the upper layers, iron, clay, and organic matter have accumulated and become cemented into a foot-thick hardpan that literally requires a jackhammer to break it apart. Plant roots cannot pass through this layer, neither can rain water. In winter, the upper soil becomes flooded and water stands on the surface; in summer, the upper soil quickly dries and roots cannot reach down below the hardpan to moist soil. In other words, not only must pygmy forest plants grow in a sterile, acidic, sandy soil beaten poor by half a million years of weather, they must also tolerate drought in summer and suffocation by standing water in winter.

Experiments have shown that dwarf plants growing on Blacklock soil are not permanently dwarfed. If planted in fertile, well-drained soil, they all grow to normal size. In fact, "pygmy cypress" trees if grown off Blacklock soil are the tallest cypress trees in California, reaching 160 feet. A better name for this conifer is "Mendocino cypress (*Cupressus pygmaea*)." Moving down the terraces, vegetation becomes progressively more luxurious with each younger terrace closer to the ocean. The third terrace supports a forest dominated by another closed-cone conifer, bishop pine. The trees are eighty feet tall and under them is a dense tangle of evergreen shrubs—the same ones found in the pygmy forest, but now growing six to ten feet tall. The soil is still acidic, but not as acid nor as low in nutrients as on the fourth terrace. No continuous hardpan exists. The second terrace

A dense grove of knobcone pine (a closed-cone conifer) amidst a fire-prone chaparral of manzanita shrubs. Most of these trees are about the same size and age because they began as seedlings immediately following the last fire.

supports a rich conifer forest of bishop pine (*Pinus muricata*), redwood, Sitka spruce, grand fir, Douglas fir, and western hemlock (*Tsuga heterophylla*) mixed with a few hardwoods such as tanbark oak and giant chinquapin (*Castanopis chrysophylla*). Underneath are many evergreen shrubs and perennial herbs. Here the soil is not deficient in nutrients, is only slightly acidic, and is completely free of a hardpan. This mixed north coast forest is among the most productive forest of the world. The juxtaposition of extremes—scrawny, scrabbling, pygmy forest and the towering, overarching redwood forest growing fewer than three miles apart on completely different soil—is typically Californian.

The Changing Landscape

Coastal forests are perhaps more diverse than forests in any other region of California. They include pygmy forests four feet tall struggling on acidic soil; ancient redwood forests of record-setting proportions; island stands of pine and cypress amid oceans of brush, which reproduce successfully only by committing group suicide in devastating fires; woodlands and savannas of rare oaks and walnuts on gently rolling hills and valleys; and old-growth mountain forests with rich mixtures of conifers, evergreen broad leaf trees, and deciduous trees.

California's coastal forests have declined markedly, and their decline parallels a statewide trend. There are now few extensive areas of old-growth forests in California. Only two centuries ago we had thirty-one million acres of primeval forest. We've lost about fifty percent of that since 1900 alone, and ninety percent since 1800. Most of what was once ancient, giant, and diverse has been reduced to so many feet of lumber and sheets of paper. Some of the land which once held old-growth forest now supports young second-growth forest of lower ecological value and different species composition, which is being intensively managed for wood and pulp.

More than a fourth of the original forest cover has been converted to other uses and it no longer supports forests of any kind. Forests dominated by redwood, Douglas fir, and ponderosa pine (*Pinus ponderosa*) have received the most pressure. Less than five percent of the original old-growth redwood forest is protected; sixty-five percent of the rest is second or third growth, and twenty-five percent has been converted to other kinds of vegetation. The mid-elevation mixed conifer forest of Douglas fir, ponderosa pine, incense cedar, and sugar pine has been reduced from fourteen million acres to about nine million acres. Just in the thirty year period 1950-1980, 125,000 acres of fir, pine, and mixed conifer—and 62,000 acres of redwood—were converted into agricultural and urban land. The California Department of Forestry predicts that 10,000 more acres of redwood, 31,000 acres of mixed evergreen forest, and 50,000 acres of fir and pine will be converted in the next thirty years, 1980-2010. That's 434 *square miles* of permanently changed landscape in sixty years.

Closed-cone conifer stands have also been lost, because much of California's development occurred along the coast. Located in the Spanish and later Mexican rancheros, most southern and central California cypress and closed-cone pine forests have been intensely grazed by cattle for a long time. Logging additionally affected some areas. Today, scattered cutting for firewood occurs in a number of Monterey, bishop, beach (*Pinus contorta* var. *contorta*), and knobcone pine stands. Greater threats to these conifer stands are road building and mining operations. Bishop pine groves in the vicinity of Lompoc in Santa Barbara County have been removed for strip-mining of diatomaceous earth. Much of the Sierra Peak Tecate cypress (*Cupressus forbesii*) grove in the Santa Ana mountains has been cut down by strip mining for clay, and the remainder is subjected to high levels of ozone pollution in smog (Chapter 5). Other conifer stands have been cleared for the removal of sand or gravel. A large sand pit endangers gowen cypress (*Cupressus goveniana*) on Huckelberry Hill in Monterey County, and mining activity for sand has cleared some stands of Monterey pine near Pebble Beach. Golf courses, urbanization, smog, resorts, seaside homes, and mountain cabins have made inroads into other stands. This replacement by development is most serious on the periphery of the mainland Torrey pine (*Pinus torreyana*) stand near San Diego and in groves around Monterey. Finally, our disruption of the natural fire cycle has also reduced the vigor of these fire-dependent species.

Our view west from atop the fog fence shows us a landscape fundamentally changed in the past two centuries. But now it's time to walk east. We've climbed to the top of the Coast Range, moving easily from deep shade of redwood up through drier, warmer air beneath madrone and pine, crunching over years-thick accumulations of discarded leaves. We crest the top ridge and gain a first distant view of the great Central Valley ahead. Summer heat there replaces summer fog. The vegetation types uniquely adapted to interior heat and aridity surround us as we drop down onto valley-facing slopes. This new vegetation is as intricate, bizarre, fragile, important, and changed by the human race as that of the fog fence.

4

VALLEY HEAT

ON A LATE SUMMER DAY 143 YEARS AGO, EDWARD BELCHER WAS aboard a small boat making its way up the Sacramento River from the delta. A captain in the British navy, he was impressed with the jungle-like wall of vegetation that lined the rivers. He wrote that the banks were:

> . . . well wooded, with oak, planes, ash, willow, walnut, poplar, and brushwood. Wild grapes in great abundance overhung the lower trees, clustering to the river, at times completely overpowering the trees on which they climbed, and producing beautiful varieties of tint. . . Within, and at the very edge of the banks, oaks of immense size were plentiful. These appeared to form a band on each side, about three hundred yards in depth . . . Several of these oaks were examined. . . The two most remarkable measured twenty-seven feet and nineteen feet in circumference, rose perpendicularly at a (computed) height of sixty feet before expanding its branches, and were truly a noble sight.

Twenty-five years later John Muir crested the Coast Range on his first walk through California. It was spring, just after his thirtieth birthday, and the sight of the Central Valley spread before him provided him with great passion and joy:

> When I first saw this central garden. . . it seemed all one sheet of plant gold, hazy and vanishing in the distance. . . Descending the eastern slopes of the Coast Range through beds of gilias and lupines. . . I at length waded out into the midst of it. All the ground was covered, not with grass and green leaves, but with radiant corollas, about ankle-deep next to the foothills, knee-deep or more five or six miles out. . . The radiant, honey-ful corollas, touching and overlapping, and rising above one another, glowed in the living light like a sunset sky—one sheet of purple and gold. . . Sauntering in any direction, hundreds of these happy sun-plants brushed against my feet at every step, and closed over them as if I were wading in liquid gold. The air was sweet with fragrance, the larks sang their blessed songs, bees stirred the lower air with their monotonous hum. . . and small bands of antelopes were almost constantly in sight, gazing curiously from some slight elevation and then bounding swiftly away with unrivaled grace of motion.

Riparian forest along the Sacramento River, dominated by Fremont cottonwoods and lianas of California grape.

Poppys and bush lupines (right) on the edge of foothill woodland in the Central Valley.

Cross-section of the western half of the Central Valley (below), as it appeared a century ago. Foothill woodland (a) leads down to savanna and finally grassland (b), then to tule marsh in the depression (c), and finally to riparian forest (d) on natural levees built from flood-borne sediments. Distances are not to scale.

The Central Valley of California, measuring sixty miles wide by 400 miles long, includes fifteen percent of the state's total area. In Captain Belcher's day, and even in Muir's, it contained a rich assortment of vegetation: wet tule marshes, tangled riparian forest, open prairies, rolling oak woodland, and dry chaparral on steep hillsides. Today, most of this vegetation has been dramatically changed by a post-gold rush human population which cleared, drained, cultivated, and built upon the land. Settlers also introduced weedy, aggressive plants from other mediterranean-climate regions of the world, and these have contributed to the displacement of vegetation native to the Central Valley.

A similar range of vegetation types, in a similar landscape, occurs along the central coast in the Salinas Valley and along the southern California coast from the Los Angeles Basin to San Diego and on into rolling hills east of there.

Although these low-elevation areas have a regional mediterranean climate,

a   b   c   d

one of the vegetation types present is an ancient relict of a summer-wet climate that existed in California more than twenty million years ago. This vegetation has been able to persist because it lives in a microenvironment that has abundant summer moisture from creeks, streams, and rivers: the riparian habitat.

The Riparian Forest

The word riparian comes from the Latin *ripa*, a stream bank or river bank, or *riparius*, growing along a bank. Riparian vegetation is the vegetation along the shores of bodies of fresh water. Along low-elevation rivers in California riparian vegetation is typically a rich forest with a mixture of genera that make Easterners feel at home: oak, maple, willow, sycamore, walnut, ash, alder. Riparian vegetation creates a diversity of wildlife habitat. Fish and amphibians thrive in the cool waters of sheltered stream banks, while a wealth of birds and mammals feed, breed, and migrate within the riparian overstory.

Approximately twenty million years ago, when the north temperate zone was wetter and less seasonal, much of North America was covered by a nearly continuous forest belt of these broadleaved species, mixed with some conifers. As mountains rose and the climate warmed and dried, the continent's center became too severe for this mesic forest. The forest segregated into two parts. It continued to have broad coverage in what today is the eastern United States, which experiences cold winters and wet, humid summers. But in drier western areas it became restricted to the riparian zone. Along these waterways the trees sink roots down to a shallow, permanent water table and thus are able to compensate for dry summer weather. Despite the fact that Central Valley winters are mild, these trees retain a cold-winter ancestral trait: they drop their leaves in winter. A winter-deciduous habit is rare in low-elevation California, yet most woody species of the riparian forest are winter-deciduous.

As a river meanders, there is an eroding bank on one side and accumulating sediment on the other. On the sediment side vegetation zones spatially replace each other. First there is sediment kept bare by the frequency and duration of flood waters. Few species are capable of tolerating continuous flooding, which prevents roots from obtaining adequate oxygen and causes poisoning of plant tissues by an

Flooding often inundates shrubby willows and cottonwoods adjacent to major rivers.

accumulation of the chemical products of incomplete respiration. Some riparian trees with tiny, wind-blown seeds require bare sand or silt on which to germinate, but they will survive only if the frequency and duration of flooding lessens— exactly the situation just a bit higher and farther from the water.

Slightly above the continuously flooded area is a shrub zone dominated by willows. The shrubs grow densely and their branches and roots trap sediment,

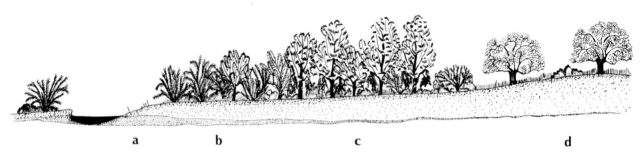

a   b   c   d

Cross section through the riparian zone. Nearest the water is an almost bare strip (a) frequently scoured by river flows. Above this is a shrub-dominated area of willows (b). Yet higher is the riparian forest (c), with several canopy layers and dominated by Fremont cottonwood. Highest and least frequently flooded, is valley oak woodland or savanna (d).

creating a natural levee. Young Fremont cottonwoods (*Populus fremontii*) may be seen growing up through the willows. Still higher, on banks five to twenty feet above the water, is the riparian forest described by Belcher. Its overstory averages eighty feet tall and is dominated by cottonwood associated with valley oak (*Quercus lobata*), white alder (*Alnus rhombifolia*), arroyo willow (*Salix lasiolepis*), and Oregon ash (*Fraxinus latifolia*). Tree density is high, averaging 100 individuals per acre. About forty percent of all the trees in this area are cottonwood. California black walnut (*Juglans hindsii*) is commonly found in this zone, but its distribution has been extended by humans, who planted it and encouraged it because of its food value. Below are layers of sub-dominant box elder (*Acer negundo* ssp. *californicum*), coyotebrush (*Baccharis pilularis* ssp. *consanguinea*), blackberries (*Rubus* ssp.) sand wild rose (*Rosa californica*), and many annual and perennial herbs. California grape (*Vitis californica*), poison oak (*Toxicodendron diversilobum*), and Dutchman's pipe (*Aristolochia californica*) climb up tree trunks and into the tangled canopy.

The cottonwood zone of the riparian forest has the most complex architecture of any California vegetation, and the richest collection of animal species. More species of birds nest in this forest, for example, than in any other California plant community. Also, twenty-five percent of California's 502 kinds of native land mammals depend on riparian habitat. Of these, twenty-one are facing threats of extinction through habitat loss.

A fourth zone, farther back from the water, is the valley oak forest or woodland. Though its overstory contains many species of the cottonwood zone, about seventy-five percent of the trees are valley oak. Valley oak is one of the largest oaks in the world, and trees can reach trunk diameters of twelve feet, heights of 150 feet, crown diameters of 150 feet, and ages of 300 to 500 years. Equally large California sycamores (*Platanus racemosa*) accompany the oaks, and these massive trees appear evenly spread in a park-like setting over a continuous ground cover of grasses and forbs. Both oak and sycamore grow best where the water table is about thirty-five feet below the surface and flooding is occasional, in contrast to alder and cottonwood, which dominate where the water table is only ten to twenty feet deep and flooding is frequent.

Cleared of trees, the coarse levee soils have proven ideal for orchards of pears, peaches, apricots, and walnuts. Some of the most expensive farmland in the state exists today where riparian forest once stood. As a result, less than ten percent of the Central Valley's original riparian cover remains. Two centuries ago the Central Valley had more than 900,000 acres of riparian forest. About 800,000 acres remained in 1848, and today less than 100,000 riparian acres still exist, more than half of which is degraded and in poor condition.

In pristine times the riparian forest extended away from the major rivers for a distance of three miles on either side. The natural levees, built up of coarse sediment deposited by annual floods, gradually dropped off in height away from the river and led into low-lying land which held flood waters or rain runoff for months at a time. These lowlands, up to five miles across on either side of rivers, supported dense stands of tules or bulrushes (*Scirpus acutus, S. californicus*) (Chapter 2). Tules are reed-like plants which reproduce vegetatively from rhizomes and send up green, leafless stems five to ten feet tall.

The original two million acres of tule marsh provided essential habitat for migratory waterfowl along the Pacific flyway. Nineteenth-century settlers found it easier to navigate this part of the Central Valley by boat than by land during winter and spring. Malaria, introduced to the west coast by Captain Cook's crew, was a serious problem in this part of California. During the twentieth century tule marshes were diked, drained, cleared, and planted with vegetables to such an extent that today only local ribbons or patches of marsh remain.

A dramatic example of the loss of marshland was the extermination of Tulare Lake. An extensive tule marsh once existed around the shores of this San Joaquin

Freshwater Tule Marsh

Winter-deciduous California sycamores and evergreen oaks form a narrow riparian zone along this small stream.

Valley lake. Nineteenth-century surveyors and settlers described it as over 700 square miles in area, reaching depths of over sixty feet. Its size varied from year to year, depending on snowpack in the Sierra Nevada. The lake was fed by the Kings, Kaweah, White, and Tule rivers and was situated in a region that otherwise received only six inches of annual precipitation. Lake waters filled a basin with no outlet. Surrounding it was a band of tules two to three miles wide. Tulare comes from the Spanish word for tules, *tulares*.

The local Yokuts traveled over the lake's surface on rafts made of tules, much as other Native Californians traveled in the San Francisco Bay area. According to historian Gerald Haslam, food was so plentiful in the area that the Yokuts had no stories of starvation in their oral history. Turtles, fish, beaver, and otters were abundant. Jedediah Smith took 1,500 pounds of beaver pelts in a single visit during the 1820s. Terrestrial animals such as antelope, bear, coyote, and elk were also common. Wildfowl were there in "multitudes," according to John Fremont during an 1844 visit. Professional hunters and fishermen sent waterfowl, fish, frogs, and turtles north to San Francisco until the turn of the century.

During the early part of the twentieth century the Homestead Act promoted settlement in the San Joaquin Valley and diversion of river water to irrigation, rather than to Tulare Lake. Dam construction in the 1930s dried the lakebed, and the surrounding tule area was converted to farmland. "The sloughs, the channels, the tules are gone," Haslam writes in his article, When Bakersfield was an Island. "So are the antelope and Yokuts who once prospered there. In fact, Tulare Basin is gone—drained, plowed, transfigured. It remains only a haunting geography of the mind. . . a ghost fading from even the possibility of memory."

Tule marsh, however, does preserve some memory of the vanished ecosystem. Soil beneath tule marsh is black from partly decomposed organic matter, and it is this organic storehouse which makes tule soils so fertile. Wherever black soil is visible between crop plants, there is a former tule marsh. One problem associated with farming organic soil is that aeration promotes decomposition, and the land surface falls. Cumulative subsidence in some delta areas now amounts to twenty feet, so that the tops of orchard trees are level with delta water, kept back only by surrounding levees. When a break in a levee occurs—as has been common every year in the delta of the San Joaquin and Sacramento rivers—flood waters rapidly fill the new lowlands and expensive pumping is required to reestablish dry land.

Grassland

The original California grassland blanketed much of the Central Valley and low elevations along the central and southern coast. It covered more than thirteen million acres and an additional nine-and-one-half million acres with an oak overstory. The original grassland supported large herds of pronghorn antelope, deer, and tule elk. Several of these animals had populations which were genetically unique to the Central Valley, suggesting a lengthy association between vegetation and grazing animals.

Geologists call the Central Valley a trough of mud—not very flattering to those who live there, but historically accurate. The grasslands that cover rolling upland slopes of the Central Valley sit upon sediments deposited as mud on the floor of ancient seas. The oldest sediments date to 140 million years ago; by 1.5 million years ago the sea had been displaced by its own sediment, and the Central Valley became dry land. Thereafter, additional sediments deposited were of material eroded from Coast Range and Sierra Nevada uplands, and of volcanic ejecta thrown down from Mount Shasta, Mount Lassen, and Sutter Buttes. Soils that have since developed from these parent materials are generally deep, brown, loamy, and rich in nutrients.

The Central Valley grassland was floristically different from the cooler, wetter northern coastal prairie. The original grassland no longer exists. What it once looked like and contained can never be known for sure, because early accounts by travelers were too vague. The best guess—and it is only a guess—is that it was dominated by two species of needlegrass (*Stipa cernua, S. pulchra*), both of which are perennial bunchgrasses with flowering stems that grow to several feet above the ground. In the open space between the clumps of grass grew annual grasses and annual and perennial forbs such as those described by Muir on his century-old trek. In spring these forbs grew lush and flowered in a spectacular show, so at that time they—not the bunchgrasses—were the dominants of the grassland. But the annuals were gone by June, and the bunchgrasses remained. Most grazing animals moved upslope, out of the grassland, in summer and returned when winter rains brought new germination and growth of the herbs.

Natural fires started by dry lightning strikes and fires purposely started by Native Californians probably burned large parts of the grasslands every year. Fire created a nutrient-rich ash and removed thatch, stimulating luxuriant regrowth of herbs the following wet season. Fire also prevented woody plants from becoming established.

Purple needlegrass, a native perennial bunchgrass that once dominated Central Valley grassland. Here it thrives on a rocky outcrop beyond the reach of domestic livestock.

All this changed in the nineteenth century. New settlers hunted the grazing animals more intensively than had native human populations. The settlers introduced domesticated livestock, keeping them at high density year-round in the valley. Some grassland was plowed and farmed. Fire was controlled. Weed seeds were introduced. In a dramatically short time the grassland was converted from a perennial cover of nutritious, native species to an annual grassland of introduced species with diminished capacity to support herds of grazing animals. The primary cause of the extinction of native grasslands was the intense, year-round grazing by livestock. The native bunchgrasses had evolved in an ecosystem with low and only seasonal grazing pressure. A bunchgrass, with all growing tips

in a cluster above ground and unprotected, is not adapted to heavy grazing. An entire plant can easily be grazed to death.

There is some controversy about the tolerance of bunch-grasses to grazing. Stephen Edwards, director of the East Bay Regional Parks District, has pointed out that the Ice Age collection of grazing-browsing-trampling animals in California grasslands was diverse and numerous, and that bunchgrasses

therefore evolved with more grazing pressure than some ecologists believe. Furthermore, bunchgrasses dominate the animal-rich east African savanna, where they are seasonally grazed to a level of only one to two inches. The key word is probably seasonal. Pristine California grasslands and modern African savannas survived only episodic or seasonal grazing. When livestock were introduced to California, the animals no longer ranged upslope off the grassland in summer, as native grazers had, but remained all year.

Cattle ranching in California began in 1769 on the coast, when the Spanish brought about 200 head of cattle to San Diego. By 1823 livestock raising was an established activity, to various degrees, at all twenty-one missions. Mission San Gabriel, near Los Angeles, owned seventeen cattle and horse ranches and fifteen for raising sheep, goats, and pigs. It is estimated that by the early 1820s San Gabriel had close to 100,000 head of cattle alone. The San Francisco Bay missions had more than 40,000 cattle, as well as other livestock. Mission Dolores used the east side of San Francisco Bay as a sheep ranch. At its peak, the missions may have had more than 400,000 cattle grazing one-sixth of California's land area.

In contrast, grazing in the Central Valley was light until the gold rush period

Native annual forbs, such as lupine and owlclover (left), produce showy grassland displays during the early spring months. In the original Central Valley grassland, these species probably clustered in patches between native bunchgrasses.

In the Central Valley grassland of today, native lupine and owlclover (right) are often scattered across a sea of non-native annual brome grass (Tehachapi Pass, Kern Co.).

California Fescue.

The standing water in this vernal pool is surrounded by concentric rings of goldfields, tidy tips, annual lupines, and owlclover (near Jolon, Monterey Co.).

after 1850. Large wild animals diminished while domestic animals rapidly increased. In 1860 the U.S. Census reported nearly one million beef cattle (not including open range cattle), just over one million sheep, and 170,000 horses and mules for all of California. As range quality declined, sheep ranching gained in favor. The number of sheep peaked at 5.7 million in 1880. By this time overgrazing and a decade of drought had taken a permanent toll on the perennial Central Valley grassland.

The second source of change in the grasslands was competition from introduced annual plants, most originating in Europe and Asia. When a perennial grass and an annual grass germinate together in a bare spot, the annual tends to win the race for domination because it produces a deeper, more extensive root system more quickly than the perennial. The annual robs the perennial of soil moisture and soil nutrients, then channels these stolen resources to its own shoot system. It produces more leaves and stems and again exploits the neighboring perennial—this time of sunlight. So while livestock were chewing back perennials, weed seeds from other parts of the world were invading the bare site created by grazing and further inhibiting the perennials.

Another factor weighing heavily against the survival of bunchgrasses is the relative palatability of native grasses compared with introduced annuals. Introduced annuals, because of their short life spans, are unavailable for grazing most of the year. Some, such as ripgut brome (*Bromus diandrus*), possess long, stiff awns on flowering stalks, which mechanically deter grazers. Others have a lower nutritional value than the native perennials. For all these reasons, domestic livestock tend to feed more intensively on native species.

Experimental fenced exclosures, which have kept livestock out of grassland for decades, typically show little recovery by native perennials. Apparently the annual weeds are such aggressive competitors and occur in such large numbers that they inhibit the return of perennials even when grazing stress is removed. Aggressive plants were inadvertently introduced by Father Serra and the first Euroamerican settlers. The numbers of both weeds and settlers have grown ever since. Today 500 weed species are well established in the Central Valley. The most common grassland dominants now are all introduced European annuals: filaree (*Erodium* spp.), soft chess (*Bromus mollis*), wild oat (*Avena fatua*), ripgut brome, ryegrass (*Lolium* spp.), foxtail (*Hordeum jubatum*), fescue (*Festuca* spp.), and California bur-clover (*Medicago polymorpha*).

Cultivation also contributed to the replacement of the original grassland with annuals. Grassland soils are excellent agricultural soils, supporting such crops as alfalfa, corn, sugar beets, cotton, tomatoes, and grapes. The largest acreage ever cultivated was during the 1880s, despite the fact that it was all dry land farming without any extensive irrigation. During the twentieth century some previously cultivated land proved unprofitable and was allowed to revert to pasture or unmanaged cover.

Certain areas within grassland are intensely, dramatically splashed with blue, yellow, and white in spring. Viewed from above, the drops of color on the landscape are shaped into circles, lines, and swirls that sometimes connect and sometimes are separate from one another. On the ground, at eye level, it is easy to detect a rolling, hillocky topography, and to see that the radiance of color comes from basins between hillocks. This is a special California place, the land of vernal pools.

The elevational difference between hillock tops and bottoms is only a few feet, but that difference turns out to be important to grassland plants. A layer of rock-hard soil, about two feet below the basins and four feet below the hillock tops, underlies these local areas. In winter the hardpan prevents water from penetrating deeper into the soil. The upper soil becomes saturated and water fills the basins, forming pools. As rainfall stops and temperatures rise in late spring, rings of vegetation left above the evaporating pool of water grow and flower, creating patches of explosive color. These basins are called vernal pools, and the plants which grow in them are called vernal pool species.

Some of the best known vernal pool species are collectively called goldfields

Islands in the Grass: Vernal Pools

(*Lasthenia* spp.)—small but intensely yellow members of the sunflower family. Other colorful and equally small vernal pool plants include meadowfoam (*Limnanthes* spp.), popcornflower (*Plagiobothrys nothofulvus*), downingia (*Downingia* spp.), butter-and-eggs (*Orthocarpus erianthus*), and checker-bloom (*Sidalcea* spp.). Less colorful but characteristic plants include coyote thistle (*Eryngium vaseyi*), annual hairgrass (*Deschampsia danthonioides*), tidy tips (*Layia fremontii*), spike rush (*Eleocharis* syn. *Heleocharis* spp.), and woolly marbles (*Psilocarphus tenellus*). All are annual herbs, and all are native species. Vernal pools have not been invaded by the introduced annuals which dominate surrounding grassland and hillock tops, probably because introduced species are not able to tolerate flooded soils low in oxygen.

Possibly all vernal pools would have disappeared by now were it not for the fact that the underlying hardpan makes the land unproductive and expensive to improve. In winter the soil is saturated, leaving crop roots without sufficient oxygen, and in summer the shallow topsoil dries, leaving crop roots stranded, unable to penetrate to deeper, wetter, layers. To farm such land, hillocks and basins must be leveled, then the land ripped by a massive clawed rig dragged behind a large tractor. As many as three passes, in different directions, are necessary to disrupt the hardpan. To plant an orchard, the entire area need not be ripped, but a hole must be blasted for every sapling.

Why hardpan forms in some soils and not in others is poorly understood, but a long passage of time seems to be required. Vernal pool soils such as those beneath pygmy forest are among the oldest in the world—up to 600,000 years old, in contrast to most soils which are 20,000 years of age or younger. During this long period of time, clay, iron, sand (silica), and organic material are leached down into the soil profile, eventually cementing together. Some hardpans consist primarily of clay and then are best called claypans, while others are constructed of iron and silica. Alkali pools in Solano and Merced counties have hardpans made of lime and silica, and pools on the Mesa de Colorado in Riverside County owe their existence to a shallow soil over a hard volcanic parent material.

Every Central Valley county contains vernal pools, but they are most abundant in Fresno, Madera, Merced, Placer, Sacramento, Solano, Tehama, and Yuba counties. Pools also occur outside the Central Valley in several southern California counties and within vegetation types above grassland, such as oak woodland and chaparral. Most pools are less than 7,000 square feet in area, but a few cover tens of acres and are temporary lakes. In pristine times vernal pools may have covered one percent of the state's area, but conversion to agriculture or urbanization has reduced this by eighty percent.

The destruction of vernal pool habitat is a profound aesthetic loss to the original grassland landscape. It is also a loss for science, because vernal pools are excellent natural laboratories for the study of plant evolution. Each pool is like an island or mountain peak, surrounded by inhospitable land unsuited for plant colonization. As a result of this isolation, many vernal pool genera contain swarms of species and varieties, each restricted to a few pools in a local area. Indeed, the vegetation in each pool is probably genetically unique, representing an evolutionary result of tens of thousands of years of geologic time. Because many species are rare and endangered, vernal pools should be fully protected.

I f the citizens of California were to select a state vegetation type, as they already have selected a state flower, state bird, state mineral, and state butterfly, then foothill woodland would be a logical choice. Oaks in the foothill woodland nurtured the highest densities of Native Californians in the entire state. The chain of Spanish missions and the early network of Spanish land grants paralleled

The vivid yellow flowers of this vernal pool contrast with the deep greens of the surrounding valley oak savanna (near left). Close inspection often reveals a wealth of species, including the blossoms of meadowfoam (white) and goldfields (yellow) near the higher, outer edges of the pool (top). As the pool slowly dries, a mixture of downingia (violet and white) and tricolor monkeyflower may be found towards the pool center (middle). With additional drying the lowest, muddy bottom may be dominated by tricolor monkeyflower later in the spring (bottom).

Above the Grass: Foothill Woodland

Foothill woodland of blue oak (right) in the dry Cuyama Valley (Santa Barbara Co.). The trees are widely spaced along the edge of the stand, forming a savanna in some places. Surrounding areas with less rainfall and south-facing slopes are dominated by annual grassland and chaparral.

A remnant savanna of valley oak (below) with a grassy understory of ripgut brome and California poppy.

the distribution of oak woodlands, and later activity by miners, ranchers, farmers, and home builders also centered in what had been pristine foothill woodland and savanna. In terms of its historic importance, the large area of the state it covers, its familiarity to us all, and its unique trees, most of which are found nowhere except in California, foothill oak woodland qualifies as the state vegetation type.

A nearly continous ellipse of oak woodland surrounds the Central Valley, about 900 miles in circumference and occupying a foothill band between 300 and 3,000 feet in elevation. The lower edge of this band is a savanna, made up of large, widely spaced valley oak trees shading grassland herbs. Tree canopy covers less than thirty percent of

California buckeye (below), a drought-deciduous tree that is often found among the driest portions of a foothill woodland.

the ground, and there are fewer than twenty trees per acre. Woodland vegetation, just upslope, has thirty to sixty percent tree cover and more than sixty trees per acre. Woodland trees are small, only fifteen to forty-five feet tall and usually less than two feet in diameter. Sometimes they are crowded together with more than 200 trees per acre, but canopy cover is still open enough to support a grassland beneath.

The species of trees which dominate foothill woodland change with distance from the moist Pacific Ocean. Close to the coast, the dominant oaks are California black oak (*Quercus kelloggii*) in the north and coast live oak. Along the interior, drier foothills—those facing the Central Valley—the dominant trees are blue oak (*Quercus*

Chinese houses, an annual herb common in oak woodlands.

douglasii), interior live oak (*Quercus wislizenii* var. *wislizenii*), and foothill pine (*Pinus sabiniana*). A small tree, often growing in clusters, is California buckeye (*Aesculus californica*). Buckeye is drought-deciduous, losing its leaves in late summer, whereas blue oak is winter-deciduous. Leaves of both are soft to the touch, in contrast to the spiny, thick, leathery, evergreen leaves of interior live oak.

The leaves of most California evergreen plants, such as live oaks, share a number of traits: they have thick cuticles; the inner sandwich of tissue is also thick; they tend to be small, often spiny, and have a leathery texture which prevents them from wilting; they are rich in tannins and other chemicals offensive to herbivores; they give off little water vapor; and they conduct photosynthesis at a low rate. As a class, such leaves are called sclerophylls (hard leaves). In terms of time and energy, they are expensive for a plant to manufacture. A lot of carbohydrate must be invested in thick walls in the numerous cells which make up the tissue. Energy-rich fats and lipids must be produced for constructing thick cuticles. Certain metabolic by-products must be accumulated for herbivore defense. The compact leaf tissue prevents evaporation and loss of water, but it also slows the rate at which carbon dioxide can diffuse into the leaf. As a result, sclerophyllous plants usually grow more slowly than soft-leaved species.

Shrubs also can be either sclerophyllous or soft-leaved. Scattered shrubs characteristically make up a second canopy layer within the foothill woodland. California coffeeberry (*Rhamnus californica*), Christmas berry (*Heteromeles arbutifolia*), and manzanita (*Arctostaphylos* spp.) are all sclerophyllous. Redbud (*Cercis occidentalis*), poison oak, and squaw bush (*Rhus trilobata* var. *pilosissima*) are all soft-leaved. They all also occur in nearby chaparral. Below them are grasses and forbs from the grassland, plus a few new ones which show that the twenty inches of annual rainfall make the woodland a bit wetter than the central valley grassland. Woodland herbs include goldback fern (*Pityrogramma triangularis* var. *triangularis*), Chinese houses (*Collinsia* spp.), hedge nettle (*Stachys rigida*), buttercup (*Ranunculus* spp.), melic grass (*Melica californica, M. torreyana, M. imperfecta*), and fescue bunchgrasses (*Festuca californica, F. idahoensis, F. occidentalis, F. rubra*).

Plant and Animal Interactions

Animal diversity is higher in the foothill woodland than in adjacent grassland and conifer forest. More than 100 species of birds live in woodlands during breeding season, and sixty species of mammals use oaks in some way for feeding, nesting, or perching. Blacktail deer rely heavily on oaks as food—eating acorns in fall and leaves in spring. The species of foothill oaks most favored by deer are California black oak, blue oak, and interior live oak. Deer also feed on the lichens which grow on lower limbs. Acorn woodpeckers use oaks for nest sites, perching, and foraging. They eat acorns and make holes in the trunk to store acorns; they excavate larger holes in the trunk and eat the sap which collects in them; in spring they feed on swelling buds and flowers.

Vegetation such as oak woodland, which is heavily utilized by a variety of animals, tends to be rich in species and structurally complex. The complexity creates a variety of habitats that can be occupied by different animals. For example, the gray fox prefers young woodlands with dense trees smaller than six inches in diameter, while the titmouse prefers older woodlands with scattered trees more than twenty-four inches in diameter. Both are part of the foothill woodland, but their niches are separate.

Oaks may be on the decline. Studies in the foothills show that oak populations do not have the age distribution expected of healthy, vigorous populations. Blue oak populations show a narrow cluster of middle-aged trees with few young or old ones. If this pattern continues for another two centuries, trees that today are eighty to 100 years old would reach the limits of their natural life span and

Appearance of foothill woodland during the summer–blue oaks with a golden understory of dead annual grasses.

disappear from the land. Some other plant community would then replace the oak woodland. Prior to the 1870s the environment permitted blue oaks to regenerate successfully, but not since. What has changed?

In Monterey County, Dr. Jim Griffin of the University of California has spent decades planting oak seedlings and watching their success under a variety of conditions. He has concluded that lack of young oaks is not due to a lack of acorns. As many as twenty acorns per square foot fall beneath oak trees each year, and most of these germinate. Death comes in the next few years from several sources: inability of young roots to penetrate soils packed hard by herds of grazing animals, late summer drought for those seedlings which germinate away from shade or on south-facing slopes, browsing by pocket gophers, browsing by above-ground rodents, and browsing by deer and cattle.

Griffin marked thousands of seedlings, and only those on north-facing slopes, in partial shade, and protected by fencing from all types of grazing survived as long as six years. Everywhere else, mortality was 100 percent. Lack of reproductive success over the past 100 years thus may have been due to drier climate, deforestation, and overgrazing by gophers, deer, and cattle. John Menke, also a professor at the University of California, has more recently been testing the competitive ability of oak seedlings against introduced annual grasses. He has found that grazing animals are an important, but not the only, cause of oak decline.

Not all foothill woodland plants are edible; many are poisonous. Poisonous plants have killed cattle, horses, sheep, people, and native animals. Some cause death if as little as one percent of the animal's weight in forage is eaten. The most important, common, and dangerous of these plants include Klamath weed (*Hypericum perforatum*), tree tobacco (*Nicotiana glauca*), bracken fern (*Pteridium aquilinum* var. *pubescens*), California buckeye, several species of larkspur (*Delphinium* spp.), monkshood (*Aconitum columbianum*), death camas (*Zigadenus venenosus*), and milkweed (*Asclepias* spp.).

Some of these are introduced plants. Many contain alkaloids in all parts of the plant, and the alkaloids induce weakness, collapse, and lack of muscle control. The toxin in Klamath weed (see Chapter 2) is unique in that it causes sores only on unpigmented skin, such as around the muzzle of livestock. This irritation indirectly leads to weight loss or death because the animal is unable to feed. Bracken fern is usually avoided by stock unless other forage is in short supply, so it is not often a problem. Its poison is cumulative and causes internal bleeding. Buckeye leaves and fruits contain saponins that can induce abortion in grazing animals and humans. In addition, the nectar is poisonous to bees.

Harvesting the Woodlands

Early settlers realized that valley oaks were indicators of rich farm soil and a shallow water table; consequently, valley oak savannas and riparian forests were cut and planted to agricultural crops. Blue oaks have also been cleared from many thousands of acres of rangeland because of a mistaken notion that they suppress optimal forage production of herbs that grow beneath them. Recent research indicates that forage quantity and quality actually may be improved by blue oak cover.

Oak wood has high density, great strength, a pleasing grain, and high heat content upon burning. These traits make it well suited for use as fuel, charcoal, mine timbers, fence posts, lumber, furniture, tool handles, paneling, parquet, pallets, and cabinets. Even the sawdust is used. Recently, some sawmills in California have developed saws capable of handling short oak logs of small diameter and irregular shape. This invention permits more intensive use of oak trees for lumber.

The popularity of wood-burning stoves increased the importance of oak as firewood. At the same time, the high cost of fossil fuel has led power companies to investigate the practicality of using oaks and other plant products for energy production. It is estimated that our remaining oak woodlands produce 3.5 million dry tons of biomass per year. If oak woodlands were to be used on a sustained basis, only this annual growth could be harvested. As fuel, the annual increment would supply only about one percent of the state's annual energy requirements, so there is little future in oak biomass conversion for energy.

The negative side effects—ecosystem damage from continual harvest and air pollution from combustion—far outweigh any benefits. Wood burning releases so much carbon dioxide and particulate matter into local airsheds that many communities now strictly control the number and use of wood-burning stoves. The Environmental Protection Agency has imposed pollution standards on wood- burning stoves, and only those meeting these standards can be imported into or manufactured in the United States. Conservation efforts are being mounted by the Sierra Club in some counties to stop the cutting of all oak trees, even on private property, until regional woodlot plans are adopted by public agencies and boards.

Many large blue oaks have been cut to clear woodlands for grazing and to provide fuelwood.

Side by Side with Woodland: Chaparral and Fire

Seen from the air or from the ground, the texture of the landscape of the California foothills is varied. Each turn seems to bring something different: first woodland comes into view, then the next hillside is covered with dense shrubs; cresting a hill we are in open savanna and grassland, then we plunge down into a densely wooded canyon. And then the sequence of scenery repeats all over again. Each piece of the mosaic has its own color and texture: woodland is irregularly bumpy, dotted with gray-green foothill pine and blue-green oak; grassland is open and golden-brown; shrubland is uniformly muddy green, the vegetation so tightly intertwined and closely fitted to the topography that it is difficult to distinguish individual plants from a distance. Spanish explorers adopted the word chaparral from the Spanish *chaparro*, a low growing type of vegetation.

Chaparral covers about ten million acres of California. Many important watersheds which collect water for agricultural and urban needs are vegetated by

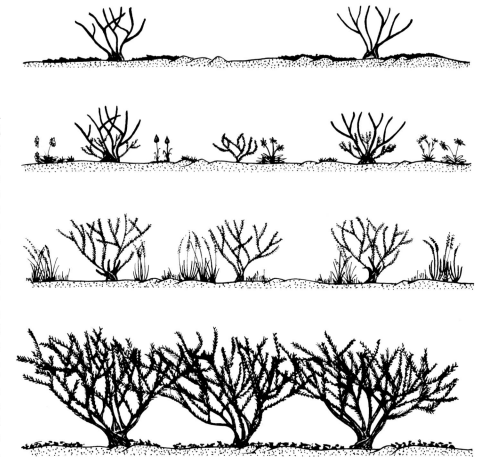

Succession in burned chaparral. Immediately after the fire (top), only skeletonized shrub canopies and ash remain above-ground. With the onset of winter rains, shrubs stump-sprout, shrub seeds germinate, and fire-following annual herbs germinate (next cross-section). Over several years, shrub canopies grow larger and the herb layer becomes dominated by grasses. Finally (bottom), after about six years, shrub canopies have regrown to their pre-fire stature and the herbs have disappeared.

chaparral. Chaparral occupies the same 300 to 3,000 foot elevation belt as grassland, savanna, and woodland. In terms of annual rainfall and the progression of seasonal temperatures, there is no difference between a chaparral site and an oak woodland or grassland; the macroenvironment or regional climate is the same. These communities coil about and mix with each other in intricate patterns because of local differences in soil depth and chemistry, frequency of fire, and slope steepness and aspect.

Typically, chaparral vegetation is a single layer of impenetrable shrubs four to eight feet tall, with evergreen, sclerophyllous leaves and rigid, intricately branched, interlacing canopies. The ground is bare of plants, stony, and covered with dry litter. An occasional California bay tree (*Umbellularia californica*), clump of cypress, or foothill pine overtops the shrub layer, and only a few perennial herbs— such as soap plant (*Chlorogalum pomeridianum*), melic grass, and globe lily (*Calochortus albus*)—fleck the ground, but most of the biomass is in a single shrub layer. Chaparral shrubs are twiggy, rather than leafy. The leaves which are present are small and vertically oriented, so the shrubs cast meager shade and lose little water. Chaparral is a drought-tolerant landscape cover.

The most common shrubs in Central Valley chaparral are chamise (*Adenostoma fasciculatum*), scrub oak (*Quercus dumosa*), Christmas berry, California coffeeberry, and more than twenty species each of manzanita and ceanothus. Several other shrubs are common in southern California chaparral: yucca (*Yucca whipplei*), red shanks (*Adenostoma sparsifolium*), laurel sumac (*Rhus laurina* syn. *Malosma laurina*), and lemonadeberry (*Rhus integrifolia*).

Chaparral occurs in a fire-prone region of California. Fires are hot, all-consuming crown fires because there may be fifty tons of dry shrubs per acre, and the shrubs accumulate flammable oils. The intensity of fire is expressed as BTUs released per second per foot of fire front. While grassland fires are relatively cool, releasing 150 BTUs per second per foot of fire front, and produce soil surface temperatures of 300°F, chaparral fires release 12,000 BTUs per second per foot in winds of only six miles per hour (much more in higher winds), and produce soil surface temperatures of 1,000°F—hot enough to melt aluminum. Chaparral fires are so intense that they create local wind storms such as occurred in the Oakland hills fire of October 1991. (It is thought that the intensity of this devastating fire was augmented by heat from burning homes and eucalyptus trees.) Fire bombs of hot air and burning wood can be hurled tens of yards in front of the fire, easily crossing fire breaks. Chaparral fires are hard to control because of these high temperatures, the leaping fire front, and the rugged terrain on which chaparral typically grows. Suburban growth in California has expanded into chaparral, exposing homeowners to potential catastrophe. Since fire is an integral part of this landscape, development of chaparral is economically unwise.

After fire, chaparral architecture changes in predictable ways. Above the ground, fires leave skeletons of shrubs and trees and a thick layer of ash. The area

looks devastated, but the promise of revegetation already lies in the soil. Seeds and roots deeper than a few inches below the surface remain alive because the insulating quality of the earth keeps them safe from the intense heat. Although temperatures high enough to melt aluminum persist for as long as twenty minutes on any surface square foot of ground during the fire, temperatures three inches below the surface reach only 150° for half that time. Roots and seeds deeper than three inches and animals in burrows are merely warmed for a few minutes. Since many chaparral shrub species have enlarged root crowns covered with dormant buds, and generally these are deep enough in the soil to survive, the shrubs have an excellent chance of surviving the fire by crown-sprouting.

With the onset of winter rains, seeds germinate, buds sprout, and roots resume growth. A carpet of herbs, shrub seedlings, and shrub sucker shoots covers much of the ground by the following spring. Most of the herbs which grow and flower this first year are not seen again until the next fire. They either have seed coats which do not permit germination unless cracked by high temperatures, or they have embryos which require some chemical released from burned chaparral wood. Until these conditions are met, seeds lie dormant in the soil, sometimes for decades.

For several years after the fire the shrub suckers and seedlings grow larger while the mix of herb species shifts in composition. Annual grasses become more numerous, while fire-following forbs eventually decline to a dormant seed pool. Within six years of the fire the shrub canopy has closed and almost all herbs have disappeared. Changes in soil chemistry and the numbers of small mammals cause the disappearance of herbs, not shade cast by growing shrubs. When leaves of chamise (and perhaps other shrubs) drop and decay, they release chemicals which inhibit the germination and root growth of most herbs, and of shrub seedlings as well. Fire removes living leaves and consumes leaf litter on the ground, eliminating the source of inhibition. The inhibitor effect does not return until the canopy closes over once again. Changes also are created by small rodents which nest beneath shrub cover. They forage for grains and young shoots most intensively near their nest sites. When chaparral burns the cover is lost and these animals move or die, releasing the site to herb growth. As the canopy closes back, animals re-enter the site and chew back the herb cover to its previous low level.

For the next thirty to forty years shrub growth continues, but at a slower pace. Woody plants which have short life spans die during this period. These short-lived shrubs are species which re-invade a burned site by seed only because they lack the capacity to stump sprout. Over time the community becomes simpler, and gradually comes to include only shrubs which sprout following fire. The dormant seeds of missing shrubs and herbs are in the soil, and will be triggered into germinating by the next fire and following winter rains. In many places chaparral fires follow each other on a twenty to twenty-five year cycle, and it is rare to find an unburned chaparral stand older than fifty years. When such old stands do burn, as they did on Mt. Diablo (Contra Costa County) in the 1970s, the firestorm is especially intense because of the high biomass of accumulated living fuel.

Chaparral vegetation can be managed if it is purposely burned more often than every twenty years, but this approach is controversial. Burned every decade, the community becomes simplified into chamisal—chaparral composed essentially of chamise shrubs only. All other species become exhausted of seed reserves or root reserves by frequent fire and finally fail to reproduce. If chamisal burns every one to two years for several years, it too reaches a point of exhaustion and fails to regenerate. At this time the site is bare and can be planted to other vegetation. Range managers typically drill in seeds of perennial grass and convert the site to grassland. The grasses used are not natives, but introduced species such as canary grass. Within two years of planting the grass cover is almost complete, and it is able to suppress any shrub seedlings which germinate from seed blown in.

The purposeful change of chaparral into grassland is called type conversion.

Stump-sprouting in chamise chaparral a few months after the fire and rain. The wavy-leaved herb is soap plant, a perennial whose bulb is protected from fire by a couple of inches of soil. A close-up (below) of manzanita sprouting from a large rootcrown or burl.

Chaparral Management

The supposed advantage of type conversion is increased access by the public for recreation, increased habitat for livestock, deer and other game animals, and improved fire control. Raging chaparral fires turn into cool, ground-hugging, slow-moving fires when they reach grassland, and the grassland also provides access for firefighting crews. Ecologically, however, the result is a major alteration of the landscape, the watershed, and natural patterns of diversity. The presence of type conversions and fuelbreaks also may promote the construction of homes in chaparral areas, which escalates the demand for yet more fire protection, illustrating how stubbornly people choose to disregard the nature of their surroundings. Intense fire is an important and predictable part of the chaparral ecosystem. On a purely cost-benefit basis, the suppression of chaparral fire probably requires more effort, money, and ongoing maintenance than is reasonable.

Here type conversion of chaparral has created expanses of annual grassland over a period of more than a decade.

If harvested, chaparral has some economic use in energy production. Detailed ecological and economic studies of chaparral's potential for biomass conversion into electrical energy have been made in southern California. A pilot project including harvest of chaparral for a generating plant near San Diego has been underway for several years with some success. Enormous mobile cutting and chipping machines have been built to mow through chaparral in a single pass and reduce it to wood chips that are trucked to the power plant and used in place of coal or fuel oil. Removing shrub tops by cutting will release dormant buds for sprouting, just as fire does. However, there is no ash to promote regrowth nor high temperatures to stimulate seed germination. We should not, therefore, expect chaparral harvested for wood to recover as fast or as completely as it does from fire.

Type conversion and biomass conversion are both controversial management options. Chaparral conveys stability to the steep hillsides on which it typically grows. Its presence no doubt affects the hydrology of watersheds, in this way having an enormous impact on water supplies for the human-made landscapes downslope. Its resilience to fire assures that vegetation recovers quickly from the natural fires which predictably come—a fire regime which cannot be entirely controlled by humans. And the natural mosaic of scrub with woodland and grassland already provides the habitat diversity required for native animals. The ecological repercussions from disturbance of chaparral for biomass harvest or type conversion should be weighed with the real economic costs and assumed benefits.

The Changing Landscape

More than thirty-one million acres today are farmed or grazed, an area equaling nearly one-third of the state. Crops have replaced large amounts of certain plant cover types. For example, rice has replaced tule marsh, and orchards have replaced riparian forest. Grazing has altered millions of acres of native grassland and woodland.

Farming and grazing give us abundant food, but modern agriculture has created toxic conditions affecting surrounding lands. For example, cotton is widely grown in the southern San Joaquin Valley in a part of the valley that receives less than six inches of rain per year. In order to grow cotton huge amounts of water with dissolved salts are imported from the delta of the San Joaquin and Sacramento rivers. Cotton farms also use large amounts of pesticides and herbicides. Those chemicals, and salts from the soil, are leached by percolating irrigation water which passes through the soil and is eventually removed from the farm in a system of drainage tiles, ditches, and canals. Over time the concentration of salts and chemicals builds up and reaches toxic levels in the drain water. Because it is toxic, the drain water cannot be reused. Early plans were to send the salt- and chemical-laden water back to the Sacramento and San Joaquin River Delta via the San Luis Drain. Fortunately, the San Luis Drain was never completed, but the waste water now drains into the infamous Kesterson National Wildlife Refuge. One of the toxic chemicals, selenium, has killed and malformed large numbers of waterfowl living in the refuge.

Soil erosion is another serious consequence of farming and grazing. It affects almost two million acres of cropland and seven million acres of grazing land. Much of the affected land is eroding at rates that will not allow sustained use of the land. In Kern County alone 17,900 acres of cropland and 526,000 acres of grazing land suffer erosion by water. Wind is eroding another 250,000 acres of cropland and 446,000 acres of grazing land. Peat soils of agricultural islands in the delta of the Sacramento and San Joaquin rivers are decomposing, blowing away, and sinking. The land has subsided in places below the foundations of protecting levees, and catastrophic flooding often results.

Agriculture has increased the number of acres of saline soils, although naturally occurring saline soils have always existed in California. Some native vegetation can grow on such soil, but most cultivated crops are too sensitive and die back if there is more than 0.5 percent salt in the soil (Chapter 2). Salinization now affects at least 1.6 million acres of cropland, mostly in the Imperial and San Joaquin valleys.

Rain, streams, and ground water carry significant amounts of minerals, and when such water is applied for irrigation, it is a two-edged sword: water comes to otherwise dry land, but salt comes with it. When the water evaporates or passes through the soil, traces of minerals are left behind. Low rainfall, high rates of evaporation, frequent irrigation, and clayey soils that retain salts are all factors that promote salinization. Salts prevent water from flowing from the soil into plants, and plants wilt from physiological drought (Chapter 2). Installing drains and using even more irrigation water to flush the soil can help reverse salinization, but this results in a larger demand for water and the creation of yet more non-reusable waste water.

Urbanization is also reducing agricultural land supply and productivity. From

1950 to 1980 more than one million acres of agricultural land were urbanized. About half of this occurred during the 1970s at a rate of 44,000 acres per year. This is twice the rate of conversion from 1950 to 1970. In five years alone—1977 to 1982—the southern California counties of Los Angeles, Orange, San Bernardino,

and Riverside urbanized 100,000 agricultural acres; San Diego County converted 60,000 acres; the San Francisco Bay area counties of Contra Costa, Alameda, Santa Clara, San Mateo, and Marin lost 41,000 acres; and in the San Joaquin Valley 65,000 acres were converted to urban uses. Another one million acres of agricultural land will most likely be urbanized between 1992 the year 2002.

Most of low-elevation California has been irreparably changed during the past two centuries. Expanses of bunch-grass prairie, vernal pools, tule marsh, and riparian forest are gone; wispy relicts hint at what once was dominant. Gone with them are tule elk, pronghorn antelope, and grizzly bear. It is no longer the same view that so moved John Muir.

The consequences of loss and change radiate out to the larger ecosystem. The shade of riparian trees, for example, provides many direct and indirect benefits to fish. Streambank cover reduces erosion and sedimentation and controls the release of nutrients to the aquatic environment. Overhanging tree canopies prevent the water from warming and losing its dissolved oxygen. Streambank vegetation also provides habitat for numerous invertebrates that are food for aquatic and terrestrial life.

This farmland near Covelo (Mendocino Co.) was developed on the rich, deep soils once dominated by oak woodland and riparian forest. It is likely that these remnant oaks will grow old and die rather than replenish the landscape with young trees.

There are complex bonds between vegetation and streams. When vegetative cover is removed, soil erosion is increased. Soil particles wash into streams, settling and clogging the stream bed. The sediment reduces spawning habitat for fish and increases the chance of flood. The effects of soil erosion can be long lasting. Hydraulic gold mining from 1850 to 1875 produced large sediment loads, and for a hundred years afterwards that sediment was deposited in drainages far removed from the diggings.

Where the riparian zone has been removed, or scarcely exists, fertilizer and animal wastes can seep into streams and ground water. An intact riparian zone acts as a filter between streams and agriculture, removing large quantities of nitrogen and phosphorus. Healthy riparian cover is the starting point of sound watershed management.

Whether in broad valleys or narrow canyons, riparian forests are among the world's most productive natural ecosystems, and among the richest in wildlife diversity. With clearing or degradation of riparian plant cover, the natural riparian ecosystem ceases to exist. Rigorous protection should be given to riparian vegetation because of its importance as a filter, its vulnerability, and its threatened extinction.

The cost of California's economic and cultural growth has been the loss and alteration of natural plant cover. If this trade of biotic resource for growth continues, we may match the history of countries such as Italy, Greece, and Israel which line the Mediterranean rim. The cradle of western history and culture surrounding the Mediterranean Sea once nestled among forests of pine, evergreen oak, laurel, ash, and myrtle. That cover has all been harvested and replaced with a modern landscape of degraded scrub. Ecologist and historian J.V. Thirgood saw a moral here for North America, a lesson which applies well to California:

> It can be argued that environmental ruin was the price paid for the glory that was Greece. . . No other part of the world so strikingly drives home the story of man's failure to maintain his environment. . . Here may be seen the fate of the newly opened lands of the new continents if man fails to achieve a balanced relationship with the land—an equilibrium between use of the land and human practice and culture.

5

CALIFORNIA'S SPINE

ONE WARM AUGUST DAY IN 1911, AMERICAN PHILOSOPHER George Santayana gave a speech on the University of California's Berkeley campus. On his first trip to California, he was impressed by its uplands. He said to his audience:

When you escape, as you love to do, to your forests and your sierras, I am sure that you do not feel you made them, or that they were made for you. From what, indeed, does the society of nature liberate you, that you find it so sweet? These primeval solitudes suspend your forced sense of your own importance. They allow you, in one happy moment, at once to play and to worship, to take yourselves simply, humbly, for what you are, and to salute the wild, indifferent, noncensorious infinity of nature.

Santayana was right. We do seem to seek the uplands, as though they fill some need unmet by daily life in the lowlands. There are even religious overtones, felt by atheists and believers alike. Sierran air, wrote Mark Twain, "is very pure and fine, bracing and delicious. And why shouldn't it be?—it is the same the angels breathe." Do we imagine that each zone of higher vegetation brings us closer to heaven, as well as to the sun and stars?

From the standpoint of vegetation, every 1,000-foot climb in elevation is equivalent to moving a distance 300 miles north. Increasing elevation brings with it lower temperatures, increasing precipitation, shallower soils, lower concentrations of oxygen and carbon dioxide, higher winds, and sometimes more unstable geology. These changes are gradual, and so are the changes in vegetation which accompany them.

There are four principal climatic zones in the California mountains: lower montane, upper montane, subalpine, and alpine. All the zones rise in elevation as one travels south toward the equator. As one range gives way to another, some species are left behind and others become more important. Direction of the mountain slope is also important. Zones shift lower on north- and east-facing aspects than they do on south- and west-facing slopes. Keep in mind always that the zones don't meet each other in narrow lines that one can walk over in a few steps. Ecotones—where zones mingle—are broad and reflect local environments more than regional ones.

Lodgepole pine forest in the upper montane zone near the Palisade Crest, Sierra Nevada (Inyo Co.)

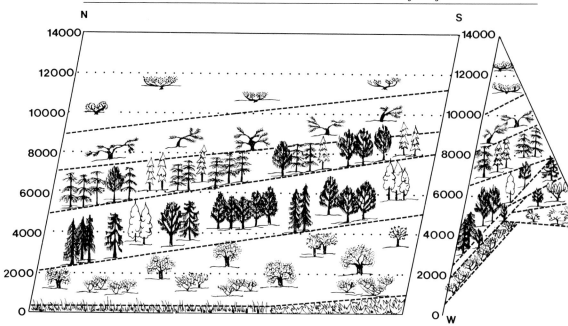

Forests of the Lower Montane

The elevations of vegetation zones in the Sierra Nevada vary from north to south and from west-facing to east-facing slopes. The same zones are higher in the south because warm, dry conditions extend farther upslope than in the north. Western slopes change from glassland to foothill woodland, lower montane forests, upper montane forests, subalpine woodland, and alpine zones. Eastern slopes change from alpine at the top to upper montane forest, Jeffrey pine forest, pinyon-juniper woodland, and desert scrub zones.

The lower montane zone is not one continuous, monotonous forest. It is a mosaic of oak-filled canyons, brushy ridges, meadows in wet flats, riparian woods, and conifer-forested slopes. Six conifers coexist and shift in importance from stand to stand: ponderosa pine (*Pinus ponderosa*), sugar pine (*Pinus lambertiana*), white fir (*Abies concolor*), Douglas-fir (*Pseudotsuga menziesii*), incense cedar (*Calocedrus decurrens*), and the Sierra bigtree or giant sequoia (*Sequoiadendron giganteum*). Forests with this rich assortment of trees are called mixed conifer forests and are extensively cut for lumber.

Ponderosa pine is the most abundant of the half-dozen trees. It is one of the most widespread conifers in North America, dominating the lower montane zone of the Rocky Mountains, the eastern slopes of the Cascades, and most of California's mountains. This is the forest John Muir described in 1894 as:

> [among] the grandest and most beautiful in the world. . . The giant pines, and firs, and Sequoias hold their arms open to the sunlight, rising above one another on the mountain benches. . . The inviting openness of the Sierra woods is one of their most distinguishing characteristics. The trees of all the species stand more or less apart in groves, or in small, irregular groups, enabling one to find a way nearly everywhere, along sunny colonnades and through openings that have a smooth, park-like surface, strewn with brown needles and burs.

Clarence King, a contemporary of Muir, was a young geologist with the Brewer survey party when he first entered the Sierra Nevada on horseback from open woodland below. He felt:

*. . . [as though you enter] a door, and ride into a vast covered hall.
You are never tired of gazing down long vistas, where, in stately
groups, stand tall shafts of pine. Columns they are, each with its own
characteristic tinting and finish. . . Here and there are wide open
spaces around which the trees group themselves in majestic ranks.*

The forests which Muir and King saw are rare today because of logging and
fire suppression. Half the original acreage of the mixed conifer forest has been clear
cut or select cut at least once in the last 150 years. The overstory conifers once
averaged more than three feet across and 100 feet in height. Although the trees
were big, they were loosely grouped, with spaces between populated by groups of
younger trees, and so tree canopies were not continuous. Probably only half the
ground was shaded.

Scattered through the forest then and now is an open understory of deciduous trees twenty to fifty feet tall—mainly California black oak (*Quercus kelloggii*) and mountain dogwood (*Cornus nuttallii*)—which add an element of color in the fall. Closer to the ground are patches of shrubs, such as greenleaf (*Arctostaphylos patula*) and whiteleaf (*Arctostaphylos viscida*) manzanita, deerbrush (*Ceanothus integerrimus*), and mountain misery (*Chamaebatia foliosa*). In wetter areas are huckleberries (*Vaccinium* spp.) and western azaleas (*Rhododendron occidentale*).

Visible in early summer is a fourth open layer of plants, a diverse collection of perennial herbs and prostrate shrubs: lupines, violets, bunchgrasses, coral root orchid (*Corallorhiza* spp.), prince's pine (*Chimaphila menziesii*, *C. umbellata* var. *occidentalis*), snow plant (*Sarcodes sanguinea*), trail plant

*A spacious stand of
ponderosa pine near
Donner Lake (Nevada Co.).*

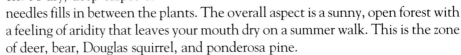

The thick evergreen leaves of greenleaf manzanita (above), with fruits.

The flowering stalk of snowplant (right), a parasitic herb often found in montane forests.

(*Adenocaulon bicolor*), squaw carpet (*Ceanothus prostratus*), pine-mat manzanita (*Arctostaphylos nevadensis*), and many others. A dry, deep carpet of needles fills in between the plants. The overall aspect is a sunny, open forest with a feeling of aridity that leaves your mouth dry on a summer walk. This is the zone of deer, bear, Douglas squirrel, and ponderosa pine.

Fire   The openness of natural lower montane forest apparently is caused by frequent ground fires. These fires consume litter, young trees, and brush, but are cool enough to leave larger trees unharmed except for an occasional scar, where temperatures get locally hot enough to burn through the bark and kill some of the cambium. In time scars are grown over and buried by succeeding rings of wood. If the tree is later cut, the number of fires survived in the life of that tree can be counted by the number of buried scars. Each scar can also be dated, by counting annual rings back from the surface.

Several extensive studies of fire scars show that fire frequency prior to 1875 averaged about one fire every eight to fifteen years in pine-dominated sites and every sixteen to thirty years in wetter fir-dominated sites. Most of these were ground fires of low intensity and limited extent; few were raging crown fires covering large areas. Some of the fires were ignited by Native Californians, some were started by lightning, but the scars don't reveal the causes.

Fire is a part of the montane environment, and its plants evolved with fire. Many plant species in this zone not only tolerate fire, they require periodic fire for optimal growth and survival. For example, ponderosa pine and giant sequoia have small seeds. When mature, these seeds drop from cones to the ground in fall and germinate the following spring. If the ground is covered by a few inches of litter, seedlings cannot extend their roots to the mineral soil below before the litter becomes too dry to support growth. Following a ground fire, the mineral soil is also fortified with nutrient-rich ash from burned vegetation. Consequently, both seedlings and mature plants are nourished by fire.

Fire also stimulates understory shrubs and herbs. Deerbrush, for example, is a common shrub in the lower montane forest. It can be killed by ground fire, but its hard-coated seeds lying in the soil are cracked and stimulated to germinate by the heat. Without the heat produced by fire, those seeds remain dormant. Fire also kills diseased or weakened trees and keeps the forest canopy open for optimal growth of understory forbs and grasses.

It is difficult to imagine the kind of ground fire that was part of the natural system, because today we see raging wildfires which consume everything in their path. Natural ground fires generate flames only one to two feet tall. Only where they consume a small tree, a clump of brush, or a downed trunk do flames jump higher. Temperatures within the low flames may reach 1,000°F, but roots and seeds within the soil are well insulated and are subjected to temperatures of only 150°F for a few minutes. Green needles on trees are scorched and turn brown only when temperatures exceed 140°F; generally needles ten feet or more above the ground escape these temperatures. The fire moves slowly, as much as three feet per second in still air, faster with a wind behind it. Wind speeds are always reduced near the ground, however, so ground fires never spread at the pace of a crown fire.

The natural place of fire in the landscape was lost sight of following disastrous fires in the late nineteenth century, when loggers, miners, and especially sheepherders torched standing timber or piles of slash on windy summer days,

Natural fires can be crown fires (left), which generate intense heat and kill all above-ground vegetation, or they can be ground fires (right), which are cooler and destroy only litter, herbs, shrubs, and saplings.

creating intense fires which swept through the crowns of acres of forests, as natural ground fires would never do. Suppression of all fires became U.S. Forest Service policy in 1905 and California Department of Forestry policy in 1924. As a result, the natural fire cycle has been disrupted for most of this century.

The absence of ground fires has had many important consequences. The balance of tree species has changed, and the forest architecture is different. With fire suppression, forests that were once dominated by trees well adapted to fire have come to be dominated by less fire-dependent trees. All conifers in the lower montane except white fir are fire-dependent. It germinates well on litter and grows successfully

Smoke rises from a low-intensity ground fire (right) as part of a prescribed burning program within Yosemite National Park (Tuolumne Co). Such fires leave the overstory largely intact.

Disastrous aftermath (below) of a crown fire within Yosemite National Park. This blaze was high-intensity and all-consuming because years of fire suppression led to high fuel loads.

in shade. Its bark is thin when young, so normally few white fir saplings would live to become overstory trees in the face of frequent ground fires. In the past sixty to eighty years, due to fire suppression, white firs have matured in vast numbers, so that the structure of

the lower montane forest has changed from an open forest to one crowded with many trees of all ages and sizes.

Not only has stand structure changed, but the likelihood of crown fire has increased due to a century's worth of accumulated dead and dried material. When fire eventually starts—and it is certain to start sometime—it will be an all-consuming, totally destructive crown fire. Flames would not stay near the ground as in the past, but would be carried up to the tallest crowns by the ladder-like arrangement of young white fir branches. The intensity of the fire would be orders of magnitude hotter because of the increased biomass on the ground and up to the canopy. The final irony is that fire suppression makes the probability of disastrous fire more likely.

The changes brought about by fire suppression can be reversed by adopting a judicious policy of prescribed burning. This policy usually requires cutting young trees first, which are piled and burned separately. Then a ground fire is set and permitted to run its course through the forest. The hand cutting of dog-hair thickets of white fir is expensive, but it may be no more expensive than the annual cost of fire suppression. Once the understory has been opened, reburning the area on an eight-year cycle is much less expensive. For the past decade, the U.S. Forest Service, the State Department of Forestry, the National Park Service, and the California Department of Parks and Recreation have all begun control burning programs. So far, fire has been used as a management tool on only a modest fraction of California's forests, but the results have been spectacular. One phase of the mixed conifer forest that has been successfully burned to protect its magnificent overstory trees from crown fire is the giant sequoia forest.

Sequoia Groves   Giant sequoias are the most massive organisms on earth. Their volume, multiplied by their weight, exceeds the mass of sperm whales—the largest animal alive—and probably of the largest extinct dinosaurs as well. The tree is pleasingly proportioned. Its massive, striated, spongy, rich brown trunk barely tapers to the first side branches that themselves bulge with great age and girth more than 100 feet above the ground. Another 100 feet above is the delightfully rounded, light green canopy, so compact that it hides all but the largest branches. The giant is all curves above, all fluted lines below. Some boles contain enormous fire scars that have repeatedly blackened a strip of trunk thirty feet tall and ten feet across. Occasionally a fallen bigtree, its roots ripped from the soil and exposed, reveals the surprisingly shallow anchor which supports the tree through centuries of winter storms. Thousands of cones litter the ground, their small size and tiny seeds a surprising contrast to the enormous size of the adult trees, which reproduce them in a chain of generations that has extended for tens of millions of years.

Neighboring trees are also large, so that it is difficult to fully grasp the enormity of giant sequoia. Sugar pine, white fir, Douglas-fir, incense cedar, and ponderosa pine all share dominance with giant sequoias. The bigtrees are so large that only three to five grow per acre. The shrubs and herbs typical of mixed conifer forest grow in scattered patches on the ground, and a thick layer of litter carpets the soil surface.

Giant sequoias are restricted to seventy-five groves, all within the Sierra Nevada, and mainly between 5,000 to 7,500 feet elevation. The groves extend along 260 miles of the west slope of the range, from Placer to Tulare

County. Some of the groves are as small as one acre and hold only six mature trees; the largest groves reach 4,000 acres and contain 20,000 trees. These groves are all we have left of a species whose fossil record extends back thirty million years. The giant sequoia was once present in such diverse areas as Europe, Idaho, and Nevada.

The first Euroamerican travelers to see giant sequoias may have been the Joseph Walker party in the Yosemite region in 1833, but that account was lost to the public. It was John Wooster, in 1850, at what is now Calaveras Big Trees State Park, and A.T. Dowd, who chased a wounded bear into the same grove two years later, who first publicized the trees. Their wild stories about enormous trees requiring twenty men with arms extended to ring the base were at first disbelieved. As word spread about these wondrous plants, a steady stream of tourists, botanists, and loggers came to see for themselves. Seeds and specimens were sent back to England, and botanist John Lindley named the tree *Wellingtonia gigantea* in honor of the winning general at the battle of Waterloo.

The genus name was changed to *Sequoia* when it was realized that a close relative, coast redwood, had been described a few years earlier. Today it is generally agreed by botanists that coast redwood and the Sierra bigtree are different enough to warrant different generic names. The name *Wellingtonia* never was resurrected, however; giant sequoia is called *Sequoiadendron giganteum*. Southern Miwoks call the tree *wawona*.

Lumbering began in 1852, and it was wasteful by today's standards. Giant sequoia wood is brittle, and the massive trunks shatter upon falling, permitting only a small percentage of the wood to be taken to sawmills. Fortunately, protection efforts began early and were largely successful. Today more than ninety percent of the remaining sequoia forests have protected status and are publicly owned. Only one-third of the original acreage has ever been harvested.

Apart from giant sequoia, there are no plant species unique to these groves, yet the massive size of sequoias visually sets them apart as special places. With a life span of 3,200 years, sequoias are not the oldest living things, nor are they the tallest, but they are the most massive. The trunk of the General Sherman tree is thirty-two feet across, rises 272 feet, and is estimated to weigh 6,000 tons. Sequoias are fast-growing, able to reach twenty feet or more across at breast height in 1,000 years. They differ from most other trees simply by virtue of growing for a longer time. Some metabolic byproducts which accumulate in the wood are natural fungicides and insecticides, and these prevent the entry of many pathogenic organisms, contributing to the tree's longevity.

A twenty-year research project on sequoias in Kings Canyon National Park, completed by four biologists from San Jose State University in 1980, elegantly showed the sequoia's fire dependency. Prescribed burns increased the number of seedlings per acre from virtually zero to 22,000. The increase resulted from several factors. First, sequoia cones do not open and release their seeds when mature;

instead they remain green, hanging onto the parent tree and retaining viable seeds for as many as nineteen years. Hot air from a ground fire causes cones to dry out, open their scales, and release their seeds: a rain of eight million seeds per acre falls after a fire. Sequoia cones also open if eaten by squirrels or beetles, but the seed rain is ten to 100 times greater after a fire.

Survival of seedlings which germinate on litter also is low. Sequoia seeds are very small: it takes 85,000 to make a pound. There is not enough food reserve in such a seed to permit seedling growth through a layer of dry, sterile litter. Seeds must germinate on bare soil, and bare soil is created by ground fire. Growth studies have demonstrated that young roots of sequoia seedlings are inhibited by acidic soil. Fire changes the natural acidity of the litter to near-neutral ash. Saplings also grow best in full sun or partial shade. Fire provides this condition by reducing brush cover and the number of young trees that would shade the young sequoia.

Public agencies which control sequoia groves are now aware that management cannot be passive. Merely locking away the forests from disturbance is not enough. Natural or planned ground fires must be permitted to sweep these groves at least once every twenty years. However, fire suppression for the past six to eight decades has permitted a dense understory of white fir to become established, and before the first control fire can be set, this fuel must be cut and burned in piles at an expense of several hundred dollars per acre. This costly type of active management is essential. Otherwise, the probability of a wildfire destroying a grove is very high. Recovery of sequoia forest from such a fire would be unlikely, because sequoias do not stump sprout as coast redwoods do. Once their tops are killed by fire, that generation—and its crop of cones and seeds—is finished. Even if young sequoias were to be hand-planted on a fire-denuded site, the enormous yet intricate structure of a mature bigtree community would not become established for thirty human generations. Control burning is now standard practice at Sequoia-Kings Canyon National Park, Yosemite National Park, and Calaveras Big Trees State Park.

Fire is not the only environmental problem in the lower montane zone. A more recently discovered one is ozone. In the 1950s foresters noted a spreading disease in the yellow pine forests of mountains around Los Angeles. Needles were flecked with yellow and their life span was shortened, so that canopies looked thin. Measurements of ring widths showed depressed wood growth. Diseased trees were more susceptible to summer drought and to attack by bark beetles, so that ultimately this disease was deadly. Ponderosa pine and Jeffrey pine (*Pinus jeffreyi*) seemed most sensitive; white fir, incense cedar, and sugar pine showed few symptoms and seemed unaffected by the disease. Since 1950 this disease—now called ozone mottle—has spread north to Sequoia-Kings Canyon National Park, Yosemite National Park, Stanislaus National Forest, and the Lake Tahoe Basin.

A grove of giant sequoias with a tall understory of fir and pine. The low understory of dogwood trees is flowering.

Air Pollution: The Invisible Killer

It has not yet been reported from the southern Cascades or the Klamath-Siskiyou-northern Coast Range area of northwestern California.

Ozone was discovered to be the cause of the disease in 1962. Ozone is a gas comprised of three atoms of oxygen. It is a highly reactive molecule, capable of deranging normal cell metabolism and structure. Formed in the lower atmosphere when automobile exhaust mixes with air in the presence of bright sunlight, it is part of the brown haze known as smog.

For more than a decade, Wayne and Joanne Williams assessed the long-term effects of air pollution on ponderosa pines in the southern Sierra Nevada mountains. They found almost two million acres of forest damaged by ozone from emissions as far as 200 miles away. Even Sequoia-Kings Canyon National Park receives enough oxidant to cause smog injury. At that time these scientists estimated that ozone damage had doubled in ten years. Smog symptoms occurred on the foliage of fourteen percent of the trees in 1975, while in 1983 twenty-four percent showed damage. Ultimately, oxidant damage affects tree growth. Trees with damaged leaves show significantly slower rates of wood growth, compared to trees with healthy leaves.

The Williams' study suggests that conifer forest decline in central California resulting from air pollution is well underway. They don't know whether these forests will be as severely affected as those in the San Bernardino and Angeles National Forests of southern California. However, production of ozone precursors is not being reduced in the Central Valley, so it is likely that Sierran forest decline will continue and expand.

If ozone damage continues for several more decades, it is likely to cause profound changes in the mixed conifer forest. The most abundant and economically important species—ponderosa pine—will gradually become less dominant. Associated conifers, shrubs, and evergreen oaks will be favored because of their higher tolerance, and their numbers will increase. If we couple this trend with that of fire suppression, then the mixed conifer forest will become a thicket of slender white fir, Douglas-fir, incense cedar, and live oak mixed and alternating with dense patches of manzanita brush. John Muir and Clarence King would not recognize it. Hikers will avoid it. Hunters and the deer they hunt will not pass through it. Lumbermen will not harvest it. But it will burn magnificently. The chaparral fires that today burn with frightening intensity do so on less fuel than will be contained in those future thickets. We have not yet seen the likes of the firestorms that may come.

Ozone is only one of many hazardous components of polluted air or smog. Other vegetation-damaging substances include a small hydrocarbon called PAN (short for peroxyacetyl nitrate), carbon monoxide, lead, nitrogen dioxide, and sulfur dioxide. Fuel processing and combustion in California annually produce more than 300 million pounds of sulfur oxides, three billion pounds of nitrogen oxides, three billion pounds of hydrocarbons, and twenty-five billion pounds of carbon monoxide. A number of pollutants injure animals and humans, as well as plants. Federal and state laws limit the maximum concentrations permissible for known health hazards such as ozone. According to the California Air Resources Board, our state's air exceeds permissible oxidant levels sixty-five days of every year. These levels are often exceeded because of the growing numbers of cars and the peculiarities of California weather and air circulation patterns, which are epitomized and accentuated in southern California coastal basins.

Along the southern California coast, cool air from the ocean frequently pushes beneath air warmed over the land. The lower layer is trapped motionless by the warm air above and by surrounding mountains. During the course of a summer day this stale air accumulates smog, which rises to the top of the trapped layer but can go no farther. At night, when temperatures cool, the trapped air can finally mix with the air above, and the smog is dissipated. Then the cycle repeats the next day. The result is a daily dense smog concentration around 5,000 feet elevation which happens to be at the lower end of the mixed conifer or yellow pine forest.

Nitrogen dioxide is a common pollutant in California smog. It combines with water to become nitric acid. Sulfur dioxide—released by smelters, the burning of coal, and certain industrial plants—is also common and forms equally corrosive sulfuric acid. California air contains more nitrogen than sulfur, but in other parts of the world sulfur is the dominant pollutant. Both gases cause rain, fog, and snow to be strongly acid.

Research by the California Air Resources Board and the California Institute of Technology shows a growing acid rain problem in the state. The acidity of any substance can be expressed as its pH—that is, as its concentration or proportion (p) of hydrogen ions (H). The smaller the pH number, the greater the acidity. Strong acids have a pH of 2 to 3; most agricultural soils have a pH of 5 to 7; normal rainwater has a pH of 5 to 6. Sulfur or nitrogen oxides cause the pH of rainwater to drop to 3 or 4. Acidity of this strength has a variety of effects on plants. It increases the solubility of some essential nutrients, permitting them to be leached from the soil and unavailable for plant roots to absorb. Acidity also increases the solubility of toxic ions such as aluminum, which can poison plant metabolism.

Acid rain damages foliage, reduces root biomass, disrupts the pattern of normal reproduction, leaches nutrients from the soil, and diminishes the tolerance of plants to frost, among a complex of many other effects that vary in intensity from one species to another. Sixteen million acres of conifer and hardwood forests in central Europe have been decimated since 1979, and most evidence points to acid rain as the cause. Destruction progresses alarmingly fast, overstory tree death following the first visual symptoms of leaf damage by only a few years. Similar symptoms are appearing in Appalachian conifer forests of the eastern United States. The possibility of similar devastation exists in California, but at present there are no symptoms of forest decline.

Soils, watersheds, and bodies of water all have some capacity to buffer acids, keeping pH constant and the ecosystem stable. Eventually, however, the natural buffering capacity is exhausted. The granitic soils of California's montane region have low buffering capacities, so the potential for a worsening acid rain problem in California is high.

Forests of the Upper Montane

Dramatic changes in vegetation occur at the upper edge of the mixed conifer forest. Most of the woody species that characterize the mixed conifer forest disappear in less than 1,000 feet of elevation gain, replaced by other plants in upper montane forests of completely different structure. One of the new conifers at this elevation is lodgepole pine (*Pinus murrayana* syn. *P. contorta* var. *murrayana*).

Lodgepole pine extends throughout the state, from extreme north to extreme south. It often grows in dense stands, especially on wet sites where snow lies late into spring or a water table is near the surface. In dry years lodgepole pine invades wet meadows, sending youngsters germinating through the opened herb cover. In wet years the water table rises and meadow plants reclaim their ground and more, encroaching beneath dying pines which suffocate in the too-wet soil. The trees are small, typically less than seventy feet tall, with trunks less than two feet across. Lodgepole pines seldom reach an age of 300 years.

At the other extreme, on dry slopes and ridges, Jeffrey pine is the dominant tree. In southern California it forms open stands alone or with white fir; in northern California it occurs mixed with red fir. Despite its dry habitat, Jeffrey pine is capable of rapid growth. Mature tree trunks commonly reach four to five feet in diameter and 160 feet in height in 400 to 500 years. Jeffrey pine is closely related to ponderosa pine, but differs in bark, cone, and leaf characters. The bark of Jeffrey pine has a rich scent of vanilla and pineapple and is more red-brown than the yellow of ponderosa. Jeffrey pine cones are twice the size of ponderosa cones and lack the outward-pointing prickles of ponderosa; they are handsomely, exceptionally symmetrical. Both yellow pines have long needles in groups of threes, but the color of Jeffrey pine needles is more blue-green than the yellow-green of ponderosa.

At the border of lower montane and upper montane, Jeffrey pine and ponderosa pine meet, mingle, and miscegenate. Although most pollen is released at different times for the two species, a small amount of pollen finds its way from one species to the other, and hybrid seeds are produced. When the hybrids germinate and mature, they possess bark, needle, and cone traits in a confusing array of combinations. Wandering among hybrids like these, the average naturalist is likely to conclude that there is no consistent difference between the two species and that the whole idea of classifying them as separate organisms is hopelessly unrealistic. Yet, when seeing pure ponderosa pine in the lower montane and pure Jeffrey pine 2,000 feet above it, the same naturalist would agree that they are dramatically different and deserve different names.

Red fir cones, like those of other true firs, grow erect on the branches.

Species usually are so genetically different from one another that they cannot interbreed and produce healthy hybrids. However, evolution is a continuing process. Young species may differ only slightly from their closest relatives. We may be seeing these pines at an early stage of separation, when they are still genetically similar enough to hybridize. It could be that, over time, the two pines will diverge further and eventually hybridization will become rare. Jeffrey and ponderosa pines are not California's only examples of hybridizing species. Oak species are notoriously promiscuous. Some evergreen oaks are able to hybridize even with nearby deciduous species. Interior live oak (*Quercus wislizenii*), for example, hybridizes with California black oak. The hybrid offspring, called oracle oak (*Quercus morehus*), has an unusual mix of leaf traits, halfway between deciduous and evergreen, halfway between soft-leaved and sclero-phyllous.

The dominant tree in the upper montane zone of northern and central California is red fir (*Abies magnifica*), one of the largest firs in the world. Its Latin name means magnificent fir, and the name is well deserved. Red fir is almost entirely limited to California, unlike its close relative white fir, which extends north to Canada and east to the Rocky Mountains.

Red fir forests are elegant and simple. Sometimes red fir is the only tree in the forest, while in other cases its companions include Jeffrey pine, lodgepole pine, and western white pine (*Pinus monticola*). Beneath the trees the stands are open. There are no scattered deciduous trees—the growing season may be too short, and

Red fir forest in the upper montane zone. The lower edge of wolf lichen growth on each trunk marks the average height of the winter snowpack.

the shade too dense—and there are few shrubs and herbs. An iridescent chartreuse growth of wolf lichen (*Letharia vulpina*) clothes the columnar fir trunks. This plant anchors itself on the rough bark of its host, but obtains no nutrition from the tree. It is a photosynthetic plant, not a parastic one, and it covers the trunk wherever light is adequate for growth. A cylinder of lichens extends down the trunk, from the canopy toward the ground. The lower edge of its growth, eight to twelve feet above the ground, approximates the average height of snowpack in winter, probably because lichen growth is not possible in the wet darkness under snow.

Snow, Shade, and the Absence of Fire

The upper montane is the zone of maximum snowfall in all of California. About eighty percent of all precipitation during the year falls as snow, and there are yards and yards of it. On average, snowpack is eight to thirteen feet deep, and it stays on the ground for 200 days of the year. In one record year seventy-five cumulative feet of snow fell at one location. The ecotone between the upper and lower montane zones is that elevation where air temperature drops below freezing during winter storms. Above this elevation, most precipitation falls as snow; below it usually falls as rain. Trees that are successful in the upper montane zone must tolerate burial under snow until they are tall enough to stand above the snowpack. Critically important features might be supple stems and mold-resistant foliage.

Even in summer, it can be cool by day beneath the shade of trees and chilly at night. The growing season is short—down from six months in the lower montane to only four months here—and the average annual temperature is a brisk 41°F.

Seedlings of red fir are fragile when it comes to withstanding heat and aridity. Plants younger than ten years of age are easily killed by the extra warmth and surface soil dryness associated with sunflecks that speckle the forest floor. Island-like clusters of fir seedlings are scattered over an ocean of bare ground. The amount of sunlight reaching the soil surface determines the presence or absence of seedlings. Only those seedlings which germinated in deepest shade survive. Red fir's requirement for a wet, cool seedbed is completely different from the requirement most mixed conifer seedlings have for open, sunny, warm habitats in the montane zone below.

Slow invasion of red fir into a 30-year-old clear cut (Stanislaus Co.).

Intolerance of full sun means that red fir does not easily invade large clear cuts harvested of timber. If clear cuts are kept narrow or small (strip cuts), then the shade cast by adjacent forest may be adequate for seedling establishment. Also, nearby red fir trees might ensure that an annual rain of seeds reaches the clear cut. Annual seed rain is important because seedling survival per year is low, so several years are required to restock a young stand. Red fir seeds also do not remain viable longer than a year. Those which fail to germinate their first spring on the ground do not last for a second spring. Fresh seeds must take their place.

A fundamental difference between lower and upper montane vegetation is that fire is not an essential element in the upper montane zone. Trees, shrubs, and herbs in this zone are neither tolerant of fire nor favored by it. Fire scars on upper montane trees have not been studied extensively enough for us to guess the average number of years between fires. However, it must be a much longer period—probably twenty-five to sixty years—compared to the eight to sixteen year cycles of the lower montane zone. Fire is a western, low-elevation phenomenon in California. There it did (and should still) occur frequently; the vegetation is adapted to it, and positive benefits follow its occurrence. However, it is a rare, destructive event from the upper montane zone east to the summits and along the desert-facing slopes of the mountains, down into the deserts themselves.

Subalpine Woodland

The dense red fir and lodgepole pine forests thin with elevation, along with the air itself. Groves of trees become restricted to patches protected from the seep of cold air and drying wind. Pruned by winter ice blast, trees become shorter, twisted, and multiple-trunked. This place of short, open forest, at the limit of tree growth, is the subalpine zone. Higher up even these bonsai-like remnants fail to grow, and the land of wood is left below. Above lies the alpine tundra.

What prevents trees from growing higher in elevation? Closer to the equater, timberline is higher than in California, so trees are not prevented from growing by thinness of the atmosphere. Nor are trees stopped by winter cold, because parts of the world far to the north are much colder, yet they support a dense conifer forest. The important limitation is warmth during the growing season: tthere isn't enough to permit a small number of leaves to feed a large, woody, perennial body. The amount of carbohydrate manufactured by photosynthesis is too close to the amount used up by respiration. The number of carbon dioxide molecules taken up by photosynthesis, divided by the number released in respiration, is too close to unity. In the temperate zone, photosynthesis exceeds respiration by a ratio of ten to one or even twenty to one, and in managed croplands it may be higher than twenty to one. But at timberline and in the alpine zone above, where the growing season is only two to three months long, daytime temperatures are cool, and frosts may occur any night, the ratio of photosynthesis to respiration is only two to one in a good year. In a poor year it may be less than 1:1. Only a plant whose shoot system is mainly green can trap enough energy and produce enough carbohydrate to sustain roots, buds, and flowers.

Few tree species are able to tolerate the special stresses of the subalpine zone. Lodgepole pine and western white pine are among the few to continue upward from the zone below, joined by mountain hemlock (*Tsuga mertensiana*) and the five-needled whitebark pine (*Pinus albicaulis*), limber pine (*Pinus flexilis*), foxtail

pine (*Pinus balfouriana*), and western bristlecone pine (*Pinus longaeva*). Not all of these trees are found together on the same mountain. Sometimes a single species will dominate; elsewhere they blend together in different ways like elements in the mixed conifer forest, and each has its own north-south limits. Of all the subalpine trees, only lodgepole pine is found all the way from Oregon to Baja California.

Not much is known about the climate of this zone. There is no U.S. Weather Bureau station at this elevation, and there are few research stations or studies of sufficient duration to provide reliable data. Average precipitation is probably about thirty inches, most of which falls as snow. There is less precipitation here than in the red fir forest below. Apparently most of the water that can be wrung out of rising air has been released before it reaches this elevation. Mean annual temperature is only a few degrees above freezing. Midday summer temperatures can be warm, but they are balanced by frequent frost at night; average temperature for the warmest month is only 50°F. Winds are high and soils are shallow and rocky, so exposed foliage tends to be sucked dry. Snow insulates buried plants against cold, protects tissues from ice blast, and keeps them out of dry air. In winter soil moisture is frozen and unavailable to roots.

Seedlings and saplings are rarely seen. Successful seedlings may require a succession of heavy snowfalls for several years, providing an insulating blanket of snow and abundant melt water in summer. They might also require active help from animals, such as the burial of seeds below ground by birds or rodents.

Once established, subalpine trees grow for centuries. Average life spans are 500 to 1,000 years, and they unfold very slowly. Foxtail pine, for example, does not begin to produce cones until the trees are older than fifty years. The trunk grows less than four inches in diameter in a century, and the needles hang onto their twigs for ten to twenty years. Old age doesn't begin to set in for foxtail pine until past the age of 1,000 years, and maximum tree age is about 3,300 years.

The subalpine zone of the nearby White Mountains contains the longest-lived trees found anywhere on earth, the western bristlecone pine. Individual trunks have had as many as 5,000 annual rings counted. In addition, dead trees on the ground rot very slowly. It has been possible to match the outermost rings of

dead trees with inner rings of those still living, so a chronology of tree rings for bristlecone pine extends back 8,000 years.

The western bristlecone tree ring record has been used to test the accuracy of the carbon-14 method of dating human artifacts. Bits of wood were taken from deep in the trunk, and their age was known by ring count. Then the wood bits were analyzed for carbon-12 and carbon-14 content and their radiocarbon age was calculated. The test revealed an error in carbon-14 dating that consistently gives dates too recent. The older the artifact, the bigger the error. Wood carbon-dated to be 3,000 years old was actually—by ring count—4,000 years old. Tree ring corrections to carbon dating have had a profound effect on the dating of archeological sites.

Subalpine woodlands can be dominated by mountain hemlock (above), white-bark pine (left, top), or western bristlecone pine (left, bottom).

Atop the Spine: the Alpine Zone

The alpine zone is a thin fringe of green near the limits of life. The growing season is measured in weeks and days, not months: six to ten weeks, forty to seventy days. The air retains little heat, so despite the sunny openness of the habitat, midday summer temperatures seldom reach 80°F, and frost occurs nightly. Plants hug the ground and sequester most of their tissue below ground, where it lies insulated from severe cold. About five times as much biomass is below ground in alpine plants as above it.

Soil moisture in summer is largely a result of snowmelt, so the densest vegetation occurs in wet meadows just downslope from late-lying snowbanks.

These meadows are rich in sedges, rushes, grasses, and herbs such as shooting star (*Dodecatheon jeffreyi*), willow-herb (*Epilobium* spp.), and veratrum (*Veratrum californicum*). Woody dwarf willows (*Salix anglorum, S. brachycarpa, S. nivalis, S. planifolia*), white heather (*Cassiope mertensiana*), and mountain heather (*Phyllodoce breweri, P. empetriformis*) are also present. More exposed ridges and talus slopes have only scattered bunchgrasses and cushion plants—perennial forbs with tiny, densely arranged leaves—such as buckwheat (*Polygonum* spp.), pussypaws (*Calyptridium umbellatum*), lupine (*Lupinus* spp.), cushion cress (*Draba lemmonii*), spreading phlox (*Phlox diffusa*), sky pilot (*Polemonium eximium*), alpine everlasting (*Antennaria alpina*), wallflower (*Erysimum perenne*), alpine daisy (*Aster alpigenus* ssp. *andersonii*), rock cress (*Arabis platysperma*), alpine buttercup (*Ranunculus eschscholtzii*), cinquefoil (*Potentilla* spp.), and alpine gold (*Hulsea algida*). Plant cover is low, but diversity is high.

The low, matted growth form of this alpine buckwheat keeps leaves and flowers close to the warm soil surface.

Alpine plants have several characteristics in common which fit them to the short growing season. Most are herbaceous perennials, a few are woody, and very few are annuals. The growing season is too short for most annuals to be able to go from seed-to-seed year after year. It is also uncommon to find seedlings of perennials. Most perennials reproduce vegetatively, by bulb or rhizome. Sexual reproduction is risky because the flowering season is short and few pollinators are available. This inhospitable alpine zone has fewer pollinating organisms than any other region in California.

Flowers of alpine plants are large and colorful. Extravagant advertising does, perhaps, attract just a few more pollinators. Some dish-shaped flowers face the sun and track it through the day, like a radar dish fixed on a satellite. The result is a little microcosm of warmth, several degrees warmer than surrounding air. The warmth is attractive to cold-blooded pollinating insects, which tend to stay longer in such flowers and to pick up more pollen as a result. Flowers open early in the growing season, only ten to twenty days after snowmelt. They can open fast because flower buds are formed one to three years before the season in which they open. Some species are capable of sending flower stalks right up through a remaining few inches of spring snow. But

Tuolumne Meadows (Tuolumne Co.), a complex of wet Sierran meadows dominated by sedges, rushes and grasses.

despite these tricks, pollination is still not assured. Eighty percent of alpine plant species are capable of self-pollination if insects, wind, or birds don't cross-pollinate them first.

Respiration proceeds faster at low temperatures in alpine plants than it does in subalpine plants. Some alpine plants, such as lungwort (*Mertensia ciliata* var. *stonatechoides*), can begin growth while still covered by snow. Their hollow stems act like glasshouses, trapping warm air rich in carbon dioxide. Green stem tissue facing inward takes up carbon dioxide by photosynthesis as rapidly as if it were midsummer.

Alpine plants are small, but they are also tenacious. Some can be dated by counting the leaf scars at the top of the root crown. A few spread outward year after year from rhizomes, and their age can be estimated by dividing the diameter of the clump by an average annual rate of spread. These methods are not exact, but they yield estimates of twenty to fifty years of age for average life spans and maximum life spans several times that long.

Herb Islands in a Woody Sea

Two hundred and fifty herb species are restricted to the California alpine zone. Bits of alpine tundra atop mountains, then, are as isolated from each other as if they were on tiny tropical islands in the vast Pacific Ocean. How did they ever reach their island peaks?

Tundra plants do not have a long fossil record, compared to montane and lowland plants. Their first fossils go back only seven to ten million years, to an island near northern Greenland, at a place today so cold it is a desert with virtually no plant life. By examining fossil deposits there and elsewhere, paleoecologists have concluded that the ancient northern tundra flora was circumboreal, predictably uniform from North America to Asia. As mountains with new tundra habitats were raised to the south, the same small group of species was able to populate them, whether the mountains were in Asia, western North America, eastern North America, or Europe.

In the Ice Age the alpine zone was depressed to such a low elevation that it was more continuous. There were mini-continents of alpine tundra, instead of islands of it. During interglacial periods, or at other times of warm, arid climate (such as the Xerothermic episode of 5,000 to 8,000 years ago), the alpine zone rose and migration from peak to peak became difficult. On some low mountains the forest zones rose clear to the top, causing an entire alpine flora to become locally extinct.

In California about forty percent of alpine plants are shared with the polar tundra or with other North American mountain chains. These are our oldest alpine species, and they are remnants of past migrations. Another fifteen percent are found only in California's alpine zone. These must have evolved in place as mountains rose to alpine heights during the past ten million years.

Surprising as it may seem, alpine plants appear to have evolved from desert plants. Although deserts and alpine areas differ in heat, they share some similarities. Habitats in both are open, with lots of sunshine. Growing seasons are short. Soils are coarse-textured and contain little available water, favoring plants with large root systems. Weather is highly unpredictable from year to year, and great diurnal fluctuations in temperature occur. So desert plants are adapted for the alpine habitat.

California's alpine plants evolved in ecosystems which did not have permanent herds of large grazing animals. When such animals are brought to the alpine zone, they overgraze the vegetation and trample the soil, causing compaction or erosion. Most alpine plants are palatable to herbivores, and their growing points are exposed above ground within easy reach of grazing animals. Alpine soils are fragile, thinly covering parent rock and loose due to repeated frost heaving. Sheep were first taken in large numbers to the mountains of California during the late nineteenth century, and they decimated montane meadows and alpine tundra like a plague of woolly locusts, as Muir called them.

Grazing by sheep, cattle, and pack horses still goes on today. Under these conditions many meadows will never recover their lost biomass, diversity, and stability. The use of public land for grazing is a firmly fixed component of today's multiple use land management ethic, but critics view grazing permit fees as unrealistically low, considering the environmental damage that results.

Relatively large flowers displayed close to the ground are typical of alpine plants. This is a perennial species of pincushion from the alpine zone of Mt. Eddy (Siskiyou Co.).

Down to the Desert

Desert-facing slopes of the Cascade Range, Warner Mountains, Sweetwater Mountains, Sierra Nevada, Transverse Ranges, and Peninsular Ranges are steep, rocky, and covered with a thinner layer of soil than west-facing slopes. The poor water-holding capacity of the soil magnifies the aridity of the eastern slope. Major tree species are mountain juniper (*Juniperus occidentalis*), white fir, quaking aspen (*Populus tremuloides*), and Jeffrey pine.

Downslope toward the desert these tree species sort themselves out differently from those on western slopes of the Sierra Nevada. The upper montaine forest is a unique mixture of red fir, white fir, Jeffrey pine, and lodgepole pine. Nearby, open stands of wind-twisted mountain juniper seem to spring full-grown out of solid ridge rock. Astonishingly, these diminutive mountain junipers have a life span equaling that of the giant sequoia.

These forests of quaking aspen east of the Sierran crest near Conway Summit (Mono Co.) are composed of clones. Each clone is a single individual that has spread horizontally with many trunks over long periods of time. Differences in fall color can be used to distinguish clones.

Quaking aspen and black cottonwood (*Populus trichocarpa* var. *trichocarpa*) cover narrrow corridors along the rocky banks of fast-moving, east-flowing creeks and short-lived snowmelt streams. Both also occur on the western face, but are more abundant on the desert flank. Their shimmering, deciduous canopies turn golden in fall, revealing naked pale green twigs and trunks which stand out against winter snow until a spring flush of new leaves completes the cycle again.

Quaking aspen is the most widespread tree species in all of North America, extending across 110 degrees of longitude and 47 degrees of latitude from Newfoundland to Alaska, from the Bering Sea to central Mexico. The seeds of aspen are small and surrounded by puffs of silky hair; they can be carried on the wind for miles in great numbers from parent trees. Aspen seedlings re-

quire open, moist sites. They usually occupy riparian habitats, but they can invade clear cuts. In clear cuts, however, conifer saplings gradually appear beneath the aspens' shade and in time the site returns to conifer forest.

Once established, aspen produces horizontal roots in additional to vertical ones. The horizontal roots bear stems and roots along their length, in this way spreading the plant vegetatively. Over many years a site can become dominated by tens or hundreds of aspen trunks, each one genetically identical to the next and all connected underground. These groups extend the life of the founder tree many times over. A single tree has a maximum life span of 200 years, but the entire population of trees (the clone) can continue expanding and reproducing for thousands of years, in time coming to occupy many acres. It is not uncommon to find several clones intermingling in one area, each the result of a different parent tree which reached the site and begat a cluster of replicate offspring. Each clone can be identified by minor differences in bark and leaf color, or by the timing of bud break or leaf drop.

Jeffrey pine is an overwhelming dominant of the east-side montane zone. It creates beautiful, open, park-like forests or woodlands with regularly spaced cinnamon-colored trunks. On the best sites tree trunks reach 130 feet tall and four feet in diameter, and the canopies can shade two-thirds of the ground. Apart from occasional patches of aspen, understory trees are rare. Shrubs cover twenty percent of the ground and have partly a cold desert flavor—sagebrush (*Artemisia tridentata*), rabbitbrush (*Chrysothamnus* spp.), and bitterbrush (*Purshia tridentata*)—and partly a montane cast—greenleaf manzanita, squaw carpet, curl-leaf mountain-mahogany (*Cercocarpus ledifolius*), sticky currant (*Ribes viscosissimum*), snowberry (*Symphoricarpos* spp.), and tobacco brush (*Ceanothus velutinus*). Squirrel tail (*Sitanion hystrix, S. jubatum*), needle grasses (*Stipa* spp.), mariposa lily (*Calochortus* spp.), lupine, and mules ears (*Wyethia* spp.) are common herbs.

This Jeffrey pine woodland has been abused over the past century. Cattle and sheep grazing were excessive and destructive, especially from the 1860s through the 1920s. Thousands of acres were heavily cut in the nineteenth and twentieth centuries to supply timbers for desert mines and the booming towns that grew up around them. Most of the Lake Tahoe Basin was clearcut at that time. In Modoc, Lassen, and Plumas counties, only trees smaller than twenty inches in diameter were left after widespread logging from 1900 to 1940. These forests have regrown, but they have not yet reached the stature, composition, and structure of the pristine forests.

Chaparral—Again   Montane chaparral covers many acres of eastern slopes. Its shrubs have low, dense branches with tough, evergreen leaves reminiscent of low-elevation chaparral. Branches of adjacent shrubs often intermingle. Some of the common species include tobacco brush, mountain whitethorn (*Ceanothus cordulatus*), huckleberry oak (*Quercus vaccinifolia*), dwarf chinquapin (*Castanopsis sempervirens*), and greenleaf manzanita. A few, such as bitter cherry (*Prunus emarginata*) and sticky currant, have soft, delicate leaves which are winter-deciduous. However, low-elevation chaparral shrubs, such as chamise (*Adenostoma fasciculatum*), are missing here.

Montane chaparral differs from low-elevation chaparral in its habitat and life cycle, as well as in the species which make it up. On rocky slopes with little soil development montane chaparral can maintain itself indefinitely, without fire or other disturbances. On better soils, however, it is only a passing phase. While montane chaparral can invade a once-forested site that has been burned, logged, or damaged by wind storm or avalanche, ultimately it will be replaced by a returning conifer forest. The succession of montane chaparral to conifer forest is slow and requires many decades. Chaparral shade improves tree sapling survival beneath, but at the same time shrub roots compete for moisture with those of young trees. Tree growth is slow until the saplings become tall enough to overtop the chaparral.

Foresters have tried to speed up succession back to forest by applying selective herbicides which kill montane chaparral shrubs but not trees. Because of ecological and human health concerns, these are no longer widely applied. The most popular brush killers are related to defoliant chemicals used in Vietnam, and their connection with fetal abnormalities in animals and humans is still unresolved. Herbicides, particularly if used in steep montane areas, can be carried into surface waters some distance away, making that water unsuitable for municipal drinking water, agricultural irrigation, and maintenance of fisheries.

Pinyon-Juniper Woodland

There are two timberlines in the mountains of California: one at high elevations, where summers are too cold to support a woody tree, and the other at low elevations, where summers are too dry. At high elevations trees give way to herbs, while at low elevations they give way to desert shrubs. The pinyon-juniper woodland forms a low-elevation timberline—the last gasp of montane trees before the unrelenting aridity and desert scrub below.

At the desert fringe water is the limiting factor. Yearly tree growth is slight. Though only fifteen to twenty feet tall, Utah juniper (*Juniperus osteosperma*) and single-leaf pinyon (*Pinus monophylla*) can reach ages of several hundred years. Despite pinyon's small stature, and the small size of its cones, its seeds are large and they are produced in great numbers. Pinyon seeds (called pine nuts) were a major food source of Native Californians (see Chapter 7). They continue to be used today by many cultures in California.

There are several related small pines—all called pinyons—which grow at the

This stand of montane chaparral is dominated by manzanita and is slowly undergoing succession to white fir forest.

desert's edge in southwestern North America. They are part of the Madro-Tertiary Geoflora (Chapter 1). California has three species. The range of Rocky Mountain two-leaf pinyon (*Pinus edulis*) just barely extends into California, in the New York Mountains and the Mid Hills of the Mojave Desert, San Bernardino County. The Sierra Juarez or Parry pinyon (*Pinus quadrifolia*), with needles typically in fours, occurs along the desert flank of the Peninsular Ranges, from Riverside County well into Baja California. It apparently hybridizes with single-leaf pinyon to produce trees with all numbers of needles between one and five.

Eastward, across a sea of shrubs, are isolated desert mountains which also have pinyon-juniper woodlands between 5,000 and 9,000 feet elevation. Those ranges that are tall enough—the Kingston, Clark, New York, White, and Panamint mountains—also have montane and subalpine woodlands above the pinyon-juniper zone. The white fir which dominates those upper-elevation forests is not the Sierran white fir (*Abies concolor* var. *iowiana*) of wet mountains to the west. It is Rocky Mountain white fir (*Abies concolor* var. *concolor*), with its own unique needle and canopy shape. It finds its way west across the Great Basin Desert on mountain

peaks, like so many vegetated stepping stones, along with limber pine and western bristlecone pine, which also extend eastward toward the Rocky Mountains and do not really belong to California.

If we move east, we leave California's spine, heading for a place that writer Mary Austin called "the land of little rain" (Chapter 6). In some sense also we leave the climate, the vegetation, and the history most people call California. Yet desert scrub may be more representative of California than any other vegetation, because it covers at least a third of the state's area. It includes some of California's most economically and ecologically valuable property—some of its most barren land and some of its most beautiful. The plants that occupy this difficult ecosystem are not only the toughest, but the most fragile.

The Changing Landscape

Our non-native society has not yet managed to change upland vegetation and ecosystems in California to the same extent that we have changed the lowlands. But significant changes have occurred. Logging has reduced acreages of pristine old-growth mixed conifer, giant sequoia, and red fir forests. Diseases such as white pine blister rust have been accidentally introduced, changing the balance of tree species in the forests. Exotic hydrocarbons, oxidants, and acids have also been added to the forest air, favoring tolerant tree species at the expense of sensitive species. Livestock were introduced to alpine and subalpine meadows, leading to loss of plant cover and severe erosion. The waterways which feed vegetation and peoples far downslope have been modified.

Most significantly, the natural regimen of ground fires has been suppressed, altering California's forest structure and increasing its flammability. The only montane forest left which still experiences natural ground fires and has the open structure that John Muir described is at the southern end of the Peninsular Range in Baja California. There, in the Parque Nacional San Pedro Martir, reached by car from a steep, rutted dirt road that climbs east from the coast highway, one can still walk through 200 square miles of park-like mixed conifer forest. But this is all that remains of unmanaged forest in the entire southwest of North America.

Perhaps the alteration of watersheds and the destination of their life-sustaining flows best illustrates the scale at which we have manipulated California's uplands. As an example of watershed manipulation, let's turn to the Owens Valley. In spring a torrent of high-Sierran snow melt cascades down the precipitous eastern face to the desert floor. This tremendous quantity of water is one of California's great resources. In the natural environment it feeds desert and riparian vegetation far to the east. Until the early part of this century it filled the Owens Valley and Mono Basin watersheds with runoff, creating water tables only a few feet below the surface.

Owens Valley was one of the few areas of California in which Native Californians practiced agriculture. The soil was rich, and irrigation water was abundant. During the late nineteenth century Euroamerican immigrants found the valley equally attractive. They settled there and began to raise a variety of crops. Other immigrants turned to the south coast and began to populate the Los Angeles Basin. Although the two areas are more than 200 miles apart, they became intimately interconnected.

Few permanent rivers and major aquifers occur in the Los Angeles Basin. If population were limited by local water resources, that limit of a quarter of a million people would have been reached in 1920. But Los Angeles citizens circumvented that limit by approving a bond issue in 1905 that authorized city officials to purchase water rights and land in Owens Valley. William Mulholland, self-taught engineer, supervised construction of the first long-distance aqueduct to serve a U.S. city. The 230-mile-long aqueduct intercepts the flow of the Owens River some miles above its mouth at Owens Lake and redirects water to Los Angeles. More than 380,000 acre-feet of Sierran runoff a year leave the Owens Valley in this aqueduct. As a result of these massive diversions, Owens Lake now is a dry bed of alkali.

The flow of water was turned on for the benefit of 200,000 southlanders in 1913, stimulating such rapid population development in the Los Angeles Basin that the population multiplied six-fold by 1930. This new population's thirst exceeded the carrying capacity of "Mulholland's ditch," so another bond issue authorized construction of an extension 100 miles north of Owens Valley to tap the streams and ground water which feed Mono Lake. Mono Lake is a saline, land-locked body of water high along the eastern flank of the Sierra Nevada in a volcanic landscape. Its unique waters are rich in carbonates, salt, and brine shrimp, the latter of which feeds an enormous population of nesting wildfowl. More than ninety percent of the state's population of California gulls raise their young on islands in the lake. Since gull nests are built on the ground, eggs and chicks can be safe from predators only on islands. Grebes and phalaropes also use this area. As a result of water diversion, Mono Lake's level has fallen forty-six vertical feet, its volume has been halved, salinity has doubled, brine shrimp population has plummeted by ninety percent, island nesting sites for birds are now unprotected peninsulas, and dry lakeshore sediments rich in sulfate blow in the air, causing human respiratory problems.

Still the population of Los Angeles grows and the pace of development has not slackened. A second aqueduct was completed in 1970. Roughly eighty percent of Los Angeles water now comes via these two pipelines. Additional water comes from the Colorado River. One million acre-feet of water a year—one billion gallons a day—is transported by aqueduct from behind Parker Dam to Lake Mathews near Riverside. Yet more comes down the great Central Valley, diverted in large canals from northern California watersheds to pumping stations which lift it over the Tehachapi Mountains and down a seemingly insatiable throat. Los Angeles thus has been able to impose its growth requirements on the resources and people of an area far away.

Although Los Angeles was the first local government in California to adopt a long-distance strategy of water exploitation, it is no longer unique. Soon after Mulholland turned on the Owens Valley spigot for Los Angeles, San Francisco began work on a 150-mile-long aqueduct of tunnels and pipes stretching from the Tuolumne River watershed on the western slope of the Sierra Nevada to Crystal Springs Reservoir, just south of San Francisco.

The keystone to San Francisco's successful capture of Sierran water was a dam on the Tuolumne. By act of Congress, and in the face of considerable opposition by a young conservation movement headed by John Muir and the Sierra Club, the Tuolumne River was eventually dammed, creating Hetch Hetchy Reservoir and drowning forever a valley as magnificent as that in Yosemite National Park. Those waters now eventually reach four Bay Area counties.

The East Bay Municipal Utility District, representing other counties, also exploits western slope Sierran runoff. Several hundred million gallons a day are taken from the Mokelumne watershed to Pardee Reservoir near Stockton, then on to two Bay Area counties. Many other water diversion schemes exist. The California Water Plan is a massive manipulation of water flow which makes these three examples modest in comparison. However, at least this is a statewide plan, not a plan which serves only one local government.

Water rights law is going to be intensively scrutinized, battled over, and revised in the courts and the state capitol during the next few decades. There is not enough water for both unlimited urban/agricultural growth and for the natural ecosystem. The shrinking of Mono Lake and resulting threats to its unique biota are well known issues, but water resource specialists in Owens Valley also report dropping water tables and increasing stress on 24,000 acres of native vegetation. Owens Valley is an early indicator of conflicts in water use which pervade the entire state. Local politics can no longer permit this kind of regional impact. Whether we favor preservation of unique ecosystems that exist few other places in the world or whether we favor extended urbanization and agriculture to accommodate increased migration into California is a decision we will make within the next twenty-five years.

6

IN THE RAINSHADOW

T HE DESERTS OF CALIFORNIA LIE IN THE EVENING SHADOW OF tall mountains. A nearly continuous chain of spires and ridges, often exceeding 10,000 feet in elevation, forces the Pacific air to rise, cool, and drop its moisture on the west side of the crest. Passing east over the shattered peaks and down through sheer-walled canyons, the air no longer releases moisture to the landscapes beyond. Thus a rainshadow is created that extends well east of the tallest summits.

From eastern California to Utah's Wasatch front, rainfall is infrequent and unpredictable. Weak summer storms are dissipated by rising columns of hot, dry air. Only a few rare showers provide enough water to germinate seeds, permit photosynthesis, prevent wilting, and forestall death to vegetation. The presence of the rainshadow thus commands the lives of arid land plants and determines the unique character of California desert vegetation.

Topography and soil also exert strong influences on California desert vegetation. Where slopes and basins occur, a distinctive plant cover develops on each because of different conditions affecting plant growth. Slope vegetation is found between the steep mountainsides and the valley bottom. The undulating slope, perhaps miles wide and hundreds of feet thick, is made from the rocky debris that has washed or fallen from above. This debris pours out of the canyons and spreads onto the valley floor in great fan-shaped piles. Over long periods of time piles from adjacent canyons coalesce to produce the gentle, undulating slopes (often referred to by the Spanish word *bajada*).

Slope soils are coarse and well drained, composed of a mixture of rock, gravel, sand, and silt. The stony surface is sometimes cemented together by a natural, water-repellant glaze, forming desert pavement. During cloudbursts the rain falling on these slopes is immediately shed from the pavement as a flash flood. Channels, also known as washes or arroyos, are cut into the surface of the slope by the force of the flood. Arroyos drain away most of the water to the valley bottom, so slope soils are especially dry. Perennial plants must be exceedingly drought-tolerant to succeed in this habitat.

Basin vegetation is found on valley floors where the land surface is flat and

A salt-encrusted playa in the Carson Slough, east of Death Valley (Inyo Co.).

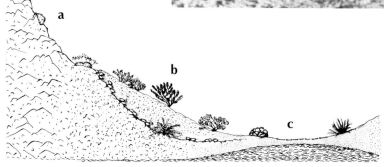

Typical mountainside (a), bajada (b) and basin (c) habitats in the desert. Mountainsides are steep, rocky, with little or no soil. The bajada is a long, gentle slope with deep, coarse soil. It is dissected by rocky drainage ways called arroyos, which are vegetated with phreatophytes. The basin has fine-textured, saline soil and a shallow water table. Distances are not to scale.

featureless and soils are rich in clay, salt, and alkali (sodium). Mountainsides and slopes provide fine material for filling and leveling the center of the valley. Shallow lakes form after a heavy storm, fed by washes that surge with water, suspended clays and silts, and dissolved salts. The lakes are short-lived, however, and evaporation leaves the clay, silt, and salts behind to form a playa, or dry lake bed.

Some playas have sodium chloride (table salt) as the predominant salt, while others also have high concentrations of calcium carbonate (lime), which make them basic in pH (the opposite of acidic). High levels of salt, lime, and alkali reduce plant growth in deserts by reducing the quality of available water and mineral nutrients (see Chapter 2). Excessive amounts of anything, even of an essential nutrient, cause plant metabolism to become unbalanced. The sodium, chloride, sulfate, and carbonate salts reach toxic concentrations in basin soils. These salts also dissolve in soil water, and plant roots can no longer absorb the moisture. Basin plants therefore experience a physiological drought. Other plant stresses in basins arise from low soil oxygen and frigid night temperatures due to drainage of cold air downward.

Geography of cold (Great Basin), warm (Mojave), and hot (Colorado) deserts (north to south) in California.

The California Deserts

Desert land in California occupies nearly twenty-eight million acres, or about twenty-eight percent of the entire state. It ranges in elevation from nearly 8,000 feet around the Mono Lake Basin to -282 feet at Badwater in Death Valley. Climatic conditions vary greatly within this range and, as a result, three distinctive types of deserts are found: cold, warm, and hot. The three desert types have been given more formal names to indicate their regional locations: the Great Basin, Mojave, and Colorado deserts, respectively.

The California portions of these deserts support a diverse assemblage of wildlife, including more than forty species of fish, sixteen species of amphibians, sixty-three species of reptiles, 420 species of birds, ninety-four species of mammals, and 1,836 species of plants. This remarkable tally excludes numerous species of fungi, moss, lichens, insects, and other invertebrates. Thus, each regional desert has a rich and unique biological identity, governed by its own climate, geology, and history. These factors are particularly important in determining the structure and species composition of desert vegetation.

Cold Desert of the Great Basin

The Great Basin, named by explorer John Fremont, is a high desert lying above 4,000 feet elevation between the Sierra-Cascade axis on the west and the Wasatch Mountains of Utah on the east. It is a composite of many small basins separated by numerous intermittent mountain ranges. From the mountains emerge rivers that carry spring runoff through steep-sided canyons. Instead of connecting to the sea, they empty into saline lakes that fill the bottoms of

Cold desert vegetation typical of slopes in the Great Basin. Bright yellow flowers of rabbitbrush (nearly five feet tall, foreground) are typical of late summer or early fall, set against the drab grey-green of basin sagebrush, saltbush and bitterbrush.

expansive valleys. Despite the presence of lakes, the climate of these valleys is that of an arid desert with warm summer and frigid winter temperatures. The Great Basin is a cold desert. In California, its northern limit is the Modoc Plateau and its southern limit is the Owens Valley. It is a land characterized by dark volcanic rock, pungent basin sagebrush (*Artemisia tridentata*), and pronghorn antelope. Stephen Trimble has called its monotonous vastness a sagebrush ocean.

The high elevations of the Great Basin experience harsh winters. The ghost town of Bodie, surrounded by sagebrush desert in eastern Mono County, is consistently one of the coldest spots in the state, with temperatures often below -10°F during December and January. Much of the yearly precipitation, usually between six and twelve inches, comes in the form of snow. After snowmelt in early spring temperatures are still too cool for plant growth. It is not until late spring, perhaps May or June, that warm temperatures coincide with ample soil moisture. By late spring, rainstorms are rare and weak and the soil moisture is quickly exhausted by the growth of herbs and shrubs. Summer drought is compounded by warm or hot days, with maximum temperatures of 80-100°F. Plants virtually stop growing then and do not begin again until rainfall once more coincides with favorable temperatures. Although the vegetative growth of Great Basin plants is restricted to late spring and early summer, flowering may occur from May to October, depending on the species and local weather conditions in any given year.

Basin vegetation in the cold desert may not be salt-affected where soil is sandy and well-drained. This lightly-grazed community is lush with Indian ricegrass, evening primrose, saltbush, and winterfat.

SLOPE VEGETATION

In its pristine form, slope vegetation of the Great Basin is a simple, two-layered mixture of shrubs, grasses, and herbs. The upper layer, perhaps three to five feet high, is composed of evergreen shrubs and tall perennial grasses that are clumped and separated by patches of open space. This community is pleasant to traverse on foot, unlike the dense, impenetrable chaparral farther west on the other side of the mountains (Chapter 4). Beneath the shrubs is a skirt of annual plants that blooms in late spring. The seeds of gilia (*Gilia* spp.), stickleaf (*Mentzelia* spp.), and many others are blown along the ground and sifted from the wind by shrubs. Their bright flowers grow up through the tangled shrub canopies and add splashes of color between the unadorned perennials.

The upper canopy layer of Great Basin slope vegetation consists of a wide variety of shrubs, although one might not realize it when passing on the highway. Woody species such as Mormon tea (*Ephedra* spp.), hop-sage (*Grayia spinosa*), cotton-thorn (*Tetradymia spinosa* var. *longispina*), and spiny menodora (*Menodora spinescens*) appear similar from a distance, but they all belong to different plant families. Mormon tea is related to pines and other gymnosperms; hop-sage is a relative of spinach and beets; cotton-thorn looks nothing like its sunflower relatives; and only the olive-like fruits of spiny menodora reveal a familial heritage. All have dense, spiny canopies of greyish green stems and small, almost inconspicuous leaves. The development of similar growth forms in different species allows them to live in similar environments. The common development of form is called *convergent* evolution.

The dominant slope species is basin sagebrush, a robust shrub with many soft, fragrant, pale green leaves. Studies of basin sagebrush in the White Mountains have shown that individual shrubs can be very long-lived, with many exceeding 800 years of age. Rabbitbrush (*Chrysothamnus* spp.) is another common shrub of the upper layer, particularly in areas disturbed by fire, grazing, or human activities such as road construction. It is a strongly scented member of the sunflower family that flowers in late summer and early fall. Desert peach (*Prunus andersonii*), a member of the rose family, produces a profusion of beautiful pink flowers in spring when other shrubs are drab with gray and green foliage. Bitterbrush (*Purshia tridentata*), another member of the rose family, is thought to be the most important food species for native browsing animals in the cold desert, particularly mule deer, pronghorn antelope, and desert bighorn sheep. The upper canopy is usually a mosaic composed of all these, and other, woody species.

Profuse pink blossums of desert peach are found in late May or early June.

Tall perennial grasses such as basin wildrye (*Elymus cinereus*), Idaho fescue (*Festuca idahoensis*), bluebunch wheatgrass (*Agropyron spicatum*), and Indian ricegrass (*Oryzopsis hymenoides*) are sometimes scattered among the shrubs. Cattle and sheep, however, have reduced or eliminated these native species from many areas. Unlike the native animals, domestic grazers prefer to eat these palatable and highly nutritious perennial grasses rather than the shrubs. Woody growth increases at the expense of the grasses, and patches of barren soil are exposed. Non-native weedy plants then readily invade the bare soil and prevent reestablishment of native perennial grasses.

One important weedy plant is an annual grass known to range managers as cheat grass (*Bromus tectorum*). Cheat grass rapidly develops a deep root system early in the growing season and exploits most of the available moisture. Seedlings of native grasses wither and die as their roots later penetrate the dry soil. Intensive grazing by cattle and sheep thus transforms slope vegetation from productive, nutritious, and perennial into low-yield, nutrient-poor, and ephemeral.

BASIN VEGETATION

On basin floors, below the remnant shorelines of ancient lakes, soils are finer in texture, poorly drained, and nutrient-poor. Any available water is extremely saline and likely to evaporate before the onset of summer drought. During the night denser cold air descends from adjacent slopes and settles in a frigid pool over the playa. Conditions for plant growth are harsh, and few species are capable of completing their life cycle. As a result, basin vegetation is dominated by a few perennial species that possess a variety of adaptations for tolerating salinity and freezing temperatures. Annual species are absent from the community because they lack adaptations to salinity and often require a frost-free growing season.

Basin vegetation is composed of grey-green, salt-tolerant shrubs arranged within a single, open canopy layer. One of these shrubs, fourwing saltbush (*Atriplex canescens*), is a highly branched species that grows to a height of three or four feet. Its fruits have papery, serrated wings that facilitate dispersal by wind. Like other species of this community, four-wing has large, hair-like glands over the surface of its leaves. The glands fill with salt that has entered the plant through the roots, eventually bursting to release brine to the outside where it dries as a crust of white crystals. The crust forms a reflective coating that reduces the amount of solar energy absorbed by the leaf and thereby prevents overheating. In contrast, the shrub known as greasewood (*Sarcobatus vermiculatus*) does not have salt glands on its bright green leaves. Instead, salt is stored in succulent tissues where it is diluted with water and isolated from sensitive physiological processes.

At elevations above 5,500 feet the slope and basin communities of the cold desert mingle with montane woodland and scrub (Chapter 5). Montane woodlands, found on slopes with well drained soil, are dominated by one-leaf pinyon (*Pinus monophylla*) and Utah juniper (*Juniperus osteosperma*). An understory of cold desert shrubs includes basin sagebrush, rabbitbrush, Indian ricegrass, and hop-sage. Slopes close to the Sierra that receive greater amounts of rain and snow (e.g., just east of Mammoth Lakes) are covered with spacious forests of Jeffrey pine (*Pinus jeffreyi*) and lodgepole pine (*Pinus murrayana*) and a basin sagebrush or bitterbrush understory. Trembling or quaking aspen (*Populus tremuloides*), a deciduous tree with bright yellow fall foliage, forms dense stands in this transitional area.

TRANSITIONAL VEGETATION

At elevations below the cold desert is another kind of desert woodland dominated by Joshua tree (*Yucca brevifolia*), with its erratic, twisted branches, dagger-shaped leaves, and fissured bark. This bizarre relative of the lily often grows to a height of thirty feet. The understory in this open woodland can be a rich mixture of cold and warm desert species. Shrubs such as fourwing saltbush, blackbush (*Coleogyne ramosissima*), hop-sage, shadscale (*Atriplex confertifolia*), and spiny menodora form an understory canopy two to three feet high. Another canopy layer is often present containing perennial grasses such as James' galleta grass (*Hilaria jamesii*) and needle-and-thread (*Stipa comata*), cacti such as old man cactus (*Opuntia erinacea*) and silver cholla (*Opuntia echinocarpa*), soft shrubs such as globemallow (*Sphaeralcea ambigua*) and encelia (*Encelia* spp.), and numerous annual species. Joshua tree woodland extends well into the Mojave desert of southern California, with extensive stands in Antelope Valley north of Los Angeles and Joshua Tree National Monument in Riverside County.

Where limestone is the dominant rock type, a transitional community rich in calcium-loving species takes the place of Joshua tree woodland. This scrub is often

the vegetation of steep mountainsides, variegated with the subtle colors of ancient seabeds: light and dark gray, burnt orange, and tan. Large shrubs that achieve dominance on other soils do not thrive on limestone outcrops where the soil is poorly developed, alkaline, and nutrient-deficient. Widespread species such as fourwing saltbush and cottonthorn are present, but they lack specialized adaptations for tolerating the soil and cannot grow and reproduce enough to dominate the vegetation.

The meager water and nutrient resources of these outcrops are thus shared by plants belonging to a greater number of species rather than by plants of a few dominant species. In the Inyo Mountains a fifty-by-fifty foot area of limestone can have twenty to thirty perennial calcium-loving species such as hedgehog cactus (*Echinocereus engelmannii*) and Newberry's locoweed (*Astragalus newberryi*), whereas an adjacent area of basalt supports only ten to fifteen widespread species. Little else is known about limestone communities of the California deserts, except that they often contain a large number of rare or endemic species. For example, July gold (*Dedeckera eurekensis*), just discovered in 1976, was considered so unusual that it constituted a new genus. Other rare plants on limestone include Mono penstemon (*Penstemon monoensis*), limestone monkeyflower (*Mimulus rupicola*), and Utah buddleja (*Buddleja utahensis*).

Communities in the transition between Great Basin and Mojave deserts: a scrub-dominated outcrop of limestone (above) in the Last Chance Mountains (Inyo Co.), with hedgehog cactus (narrow stems) and cottontop (wide stems), and Joshua tree woodland (right) with an understory of shadscale, hopsage and James' galleta grass.

The Mojave desert is an ecotone, running between the high, cold Great Basin desert of the north and the low, hot Colorado desert of the south. Its average elevation is 2,000 feet. The lower elevation creates more heat and drought than occurs in the cold desert, but winters are mild. Only four to six inches of precipitation fall in an average year, mostly between November and March. All year long temperatures are warmer than in the Great Basin, but light snow and frost still occur from December to February.

The amount and timing of the growing season's first rainfall is particularly important to plant life in the Mojave desert. Many species of annual plants germinate in late fall when an inch or more of precipitation wets the parched soil. These plants, known as winter ephemerals, spend December and January as small

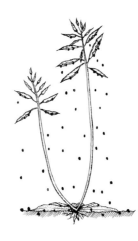

seedlings or rosettes, growing slowly, leaf by leaf, enduring the long, cold nights. They are able to take advantage of the moderate daytime temperatures (50-80°F) and rainfall of this period. Their green tissue photosynthesizes by means of the C_3 biochemical pathway common to species that grow in moist, temperate environments. Winter annuals accumulate enough carbohydrate reserves to flower, set seed, and die before onset of hot, dry summer conditions. By late May all that remains are seeds attached to a senescent, desiccated adult plant.

Shrubs of the Mojave also begin new physiogical activity in response to the first significant autumn rain, and grow primarily during late winter and early spring. This is in striking contrast to the winter dormancy and spring growth of Great Basin plants. Flowering of Mojave shrubs is also earlier than in the Great Basin, most often occurring in spring and early summer.

Other annual species, the summer ephemerals, germinate in spring or summer and grow by means of the specialized C_4 photosynthetic pathway, described in Chapter 2. Because C_4 photosynthesis allows leaves to fix carbon at high temperatures and with less water, these species are able to flower and set seed in late summer and early fall when the rest of the desert vegetation is dormant. At these times midday temperatures commonly exceed 100°F and can reach 115°F. Plants able to grow during extreme periods of heat and drought have important genes that might be bred or biologically engineered into crop plants whose thirst can no longer be met by extravagant irrigation.

Part of the Mojave desert's great need for water is quenched by the Mojave River, which begins in the San Bernardino Mountains and flows east, downcutting through magnificent Afton Canyon. Steep walls of red and yellow bedrock then open onto the expansive Devil's Playground, and the river runs among great undulations of sand for nearly thirty miles. After a hard rain the Mojave River fills several dry lake beds with mirror-like sheets of water, attracting gulls and shorebirds to an otherwise arid landscape. In rare years the river overflows to the north, where it joins the Amargosa River and placidly enters the southern portion of Death Valley.

The small volume and leisurely pace of these waters belies the great history of the river. Along with the Amargosa River, the Mojave River created Pleistocene Lake Manly, which stood 600 feet deep and covered 618 square miles in

The life history of a desert winter annual. Seeds germinate in late fall with the onset of the rainy season (left). They grow slowly during winter as a rosette of leaves close to the ground. In spring, increasing daylength and rising temperatures cause stems to bolt and flowers to bloom. Early summer drought (far right) leads to fruit maturation, seed dispersal, and death of the parent plant.

what is now Death Valley. A once mightier Mojave river carried of water for a vast system of ancient lakes and deposited enormous quantities of sand that formed Devil's Playground and Kelso Dunes. At that time the surrounding vegetation was more like that of today's cold desert, with pinyon woodlands and basin sagebrush flourishing under a cooler, wetter climate. During the last 10,000 years warmer and more arid conditions have reduced the lakes to salt-encrusted playas and led to the development of modern Mojave desert vegetation.

Traveling across the Mojave desert en route to Las Vegas, many people notice a certain regularity, even monotony, in the slope vegetation. Tall shrubs with buggy-whip stems and sparse foliage seem to be arranged as if in an orchard. Small shrubs, yuccas, and cacti form an open understory. In drier areas the spacing between plants becomes even more pronounced. Ecologists have demonstrated that competition for water between root systems can account for these patterns in the vegetation, among such other factors as allelopathy (chemical warfare between plant species), seed dispersal, and water-repellant crusts on the soil surface.

SLOPE VEGETATION

The dominant shrub of the Mojave bajada is the evergreen creosote bush (*Larrea tridentata*). With its small, hard leaves and flimsy canopy, it is hard to believe that this is the most successful desert plant in all of North America. Its success is measured in terms of distribution and longevity. Creosote bush can be found from Owens Valley, across southern portions of Nevada and Utah, south through Arizona, Baja California, New Mexico, Texas, and well into central Mexico—an area of over 274,000 square miles. Closely related species grow in the desert regions of Chile and Argentina.

Recent studies suggest that some circular clusters of creosote bush may be thousands of years old, placing them among the oldest living things on the planet. As a bush grows it accumulates a mound of wind-blown soil and debris that buries the base of its branches. Some outermost branches take root, break off from the parent, and continue growth as separate plants. Over time the original shrub will disappear, leaving a circular clone of daughter plants—a fairy ring of genetically identical offspring. Assuming that the rings expand at a slow, constant rate (perhaps 1/10 to 1/40 inch per year), a circle seventy-five feet in diameter

A well-spaced stand of creosote bush, typical of bajadas in the Mojave Desert. Silver cholla is the cactus to the left and desert pavement can be seen in the foreground.

would have originated more than 9,000 years ago. Rings this large exist in the Mojave desert. Some of the plant clusters may be daughters of the original desert colonists, invading the Mojave at the close of the ice age when a warm desert climate was first established.

Even the most ancient creosote bush is capable of producing a profusion of yellow flowers and white, fuzzy, spherical fruits each spring. Seedlings, however, are rarely observed, and new plants may become established only once every few decades, when rainfall is especially favorable.

An important reason for the success of creosote bush lies in its ability to survive extended periods of drought. Its evergreen leaves are able to maintain their basic physiological activities, including photosynthesis, after as many as thirty rainless months. Covered with a thick, waxy cuticle and constructed to withstand dehydration of internal tissues, the leaves of creosote bush conserve water when it is available and tolerate drought stress when it is not. That stress, measured as the tension (negative pressure) found in the water-conducting xylem tissues, becomes five to ten times greater in creosote bush than in crop plants or fruit trees. In extremely dry portions of the desert, high tensions can last for years. Under such conditions many species die, but creosote bush might only lose a portion of its leaves. Recovery can be rapid, marked by the production of new leaves immediately following the first good rain. If we can understand this metabolic ability to tolerate water stress, we might also understand why most crop species are so intolerant. Selective breeding or genetic engineering might then produce crop plants that physiologically mimic the creosote bush. This is a strong argument for the preservation of wild, native plants.

Other shrubs contribute to the upper canopy layer of Mojave bajada vegetation. Burro bush (*Ambrosia dumosa*) co-dominates with creosote throughout seventy percent of the warm desert's area. It has a short (one to two feet), round canopy of soft leaves that are dropped at the onset of drought. Soft leaves are capable of higher photosynthetic rates than the waxy, evergreen leaves of creosote bush, and this compensates for their short duration on the plant. Brittlebush (*Encelia farinosa*) has broader, triangular leaves that become white with dense hairs as temperatures climb and soil moisture decreases over the summer. Like the salt crust found on saltbush leaves, the whitish hairs reflect solar radiation and keep the heat load down. This structure prolongs the life of the leaf and maximizes its photosynthetic contribution to the parent plant. Indigo bush (*Dalea schottii*), another overstory shrub, is a woody member of the bean family with contorted, bone-colored stems and sparse foliage. Its unusual growth pattern produces a low, skeletal canopy that contrasts with the other, leafier species of the community. Toward the south, Mohave yucca (*Yucca schidigera*) becomes common on well drained soils.

The evergreen leaves of creosote bush are small, thick, and waxy. These yellow flowers will soon produce fuzzy, white fruits that promote wind dispersal of the seeds.

Over twenty species of cacti can be found throughout the Mojave desert. The leaves of cacti have been eliminated during evolution, and fleshy stems have assumed the job of photosynthesis. Some species, such as beavertail (*Opuntia basilaris*), have flat, pancake-like stems that ack ribs. Beavertail is one of the most widely distributed cacti and contributes to both bajada and mountainside vegetation. Others species, such as silver cholla and cottontop (*Echinocereus polycephalus*), have cylindrical stems, often with pronounced ribs displaying large spines.

Ribs act like fins on a radiator, dissipating heat on hot summer days while allowing stems to expand when water is taken up and stored in winter. Spines perform several functions. They probably keep some animals from eating the succulent tissue when water becomes scarce. A dense covering of spines also acts as a reflective coating, reducing tissue temperatures just as leaf hairs and salt coatings do in other desert species. Reflective spines and radiating ribs are particularly important to cacti, because they transfer heat from the thick tissue to the surrounding air. Most non-succulent plants release heat by allowing water vapor to escape through open stomates during the day. This loss of vapor (transpiration) is analogous to sweating in animals, and the cooling result is the same. Many cacti, however, keep their stomates closed by day and open them only at night.

These cacti possess a third unique photosynthetic pathway called CAM

Flat stems of beavertail cactus (left) are covered by clusters of fine, pungent hairs (glochids) that are difficult to see.

Stems of hedgehog cactus (right) are cylindrical and covered by clusters of large, straw-colored spines that reflect solar radiation.

(Crassulacean Acid Metabolism). In contrast to other plants that possess C_3 or C_4 photosynthesis (see Chapter 2), CAM plants are able to open their stomates at night instead of during the day. This allows for the uptake of atmospheric carbon dioxide through stomates on the stems, at just the time when water loss is minimized by cool, humid night air. Consequently, CAM plants lose less water per unit of carbon gained; that is, they are more water-use efficient. The accumulated carbon dioxide is stored as a weak acid until it can be converted into food using

the sun's energy during the day. Succulent stem tissues provide a reservoir for the acid and a source of water to sustain night-time gas exchange.

Scattered among the shrubs and cacti are delicate but showy annual wildflowers. Their abundance in any portion of the desert depends on year-to-year and seasonal variation in the amount and timing of precipitation. Dr. Janice Beatley, while working at the Nevada Test Site north of Las Vegas, determined that one inch of rain falling between September and December is enough to trigger

widespread germination of winter ephemerals. If heavy storms occur early in fall, a magnificent display of wildflowers is likely in spring. During such a year it is possible to see innumerable plants representing a great variety of species. Desert-sunflower (*Geraea canescens*), Mojave buckwheat (*Eriogonum mohavense*), gilias, Bigelow's mimulus (*Mimulus bigelovii*), desert-marigold (*Baileya pleniradiata*), brown-eyed primrose (*Camissonia claviformis*), desert five-spot (*Malvastrum rotundifolium*), Fremont's phacelia (*Phacelia fremontii*), and Shockley's lupine (*Lupinus shockleyi*) are but a few of the common winter annuals. Perhaps one year in ten, however, is favorable for a bumper crop of wildflowers. Most years bring light, sporadic rainfall in winter and spring, and winter ephemerals are sparse or absent from many areas.

Summer ephemerals germinate in response to spring or summer rains that do not stimulate winter annuals. These species are few in num-

Heavy rains in the fall promote showy springtime displays of winter annuals in the warm desert.

ber and seldom form an impressive floral display. Diminutive plants such as chinchweed (*Pectis papposa*), Wright's boerhaavia (*Boehavaria wrightii*), Yuma spurge (*Chamaesyce setiloba*), and six-weeks grama grass (*Bouteloua barbata*) grow large and showy only if summer thunderstorms provide them with extra moisture. Annuals of the bajada are therefore very dynamic, subject to and characterized by spatial and temporal variations in desert rainfall.

The uncertainty of desert rainfall requires the ephemeral component of bajada vegetation to spend most of its time in a dormant, cryptic state—the seed. A hard, waxy seed coat protects the enclosed embryo from drying during years of insufficient rainfall. Chemicals in the seed coat inhibit germination until sufficient rain leaches them away. When the right conditions exist, the embryo is activated nd begins producing leaves, stems, and roots. In effect, this mechanism gauges the rain, ensuring that available soil moisture will sustain the growth of these frail, ephemeral plants. It also means that an area covered with colorful annuals during one wet year may be devoid of them during a dry year.

Once established, annual plants display their leaves in low rosettes with minimal overlap, allowing for full illumination by the sun. Many species are capable of photosynthetic rates higher than those of our best crop plants. This enables winter annuals to reach reproductive maturity in a short time—perhaps a month—before soil moisture is depleted. Numerous flowers are then produced. In good rainfall years millions upon millions of orange, blue, white, purple, and yellow blossoms carpet the floor of the bajada. From these the next generation of seeds disperse, only to lie and wait for the next unpredictable rains.

Livestock grazing has had a significant impact on the wildflowers of the Mojave and the fragile web of wildlife they support. Desert animals are highly dependent on native annuals for food. Newly hatched desert tortoises forage for the succulent, nitrogen-rich foliage of the spring wildflowers. At this time they must accumulate most of the nutrition needed to endure their first year in the desert. Desert pocket mice and kangaroo rats cache piles of seed that will become scarce commodities in the months to come. In turn, the tortoises, rodents, insects, and other herbivores constitute potential food for desert carnivores. To a great extent, the abundance of these native animals reflects the rainfall patterns and the resultant abundance of annual plants. Livestock grazing, however, drastically reduces the amount of forage available for wildlife in wet and dry years alike. One pass of a flock of sheep can reduce the biomass of annual plants by sixty percent during the critical spring season. In 1992 the Bureau of Land Management imposed new grazing restrictions on large areas of prime desert tortoise habitat, to the relief of biologists but to the displeasure of desert stockmen.

Government data have also shown that forty to seventy percent of the annual plant production in the western Mojave is now comprised of non-native

The growth form of this annual primrose displays leaves and flowers so that self-shading is minimized. With leaves near the warm soil surface, the plant is able to sustain very high rates of photosynthesis during the cool winter months.

weeds. The weeds have spread, and their growth has been promoted by cattle and sheep, allowing them to out-compete and replace the native annuals upon which wildlife depends. Not only are food plants removed; so are the perennial shrubs needed for hiding from predators and the intense midday sun. The vegetation of an intensively grazed bajada is significantly different from that of the non-disturbed bajada: it has less plant cover, produces lower-quality forage, supports fewer native species of plants and animals, and often looks like a degraded and discarded landscape. New management perspectives and techniques will have to be deployed in order to maintain the wildflower and wildlife resources of the Mojave bajadas.

BASIN VEGETATION

The Mojave and Great Basin deserts have similar basin vegetation because the physical characteristics of their playas are virtually the same: fine soil texture, high salinity and alkalinity, scarce nutrients, frequent frost, and poor drainage. In the Mojave, however, the state of physiological drought due to salinity is accentuated by the drier, warmer climate. Salt-tolerant shrubs and small trees are found nearest the shoreline of barren lakebeds. Some, such as greasewood, inkweed (*Suaeda torreyana* var. *ramosissima*), and iodine bush (*Allenrolfea occidentalis*), are shallow-rooted and have succulent leaf or stem tissues that sequester the salt they take up. Alkali saltgrass (*Distichlis spicata* var. *stricta*) and alkali dropseed (*Sporobolus airoides*) are found as an understory component among the shrubs. Other species, especially trees such as honey mesquite (*Prosopis juliflora* var. *torreyana*), tap into subterranean water by means of a deep root up to 160 feet long. Several types of saltbush, including fourwing saltbush, desert saltbush (*Atriplex polycarpa*), and—especially in Death Valley—desert holly (*Atriplex hymenelytra*), become more prominent members of the vegetation away from the shoreline. Eventually, where the toe of adjacent alluvial slopes meets the flat basin, the soil becomes better drained and less saline, allowing for the growth of creosote bush and its typical bajada associates.

Desert basins frequently contain sand dunes. Basins fed by rivers or arroyos also are supplied with large quantities of sand washed from eroding mountains. A strong wind can gather and carry the sand until an obstruction (plant canopy, rock outcrop, or mountainside) forces it to slow down and drop the grains. These windblown accumulations may be shallow and continuous (sand sheets) or deep and concentrated (sand dunes). A well drained, non-saline soil is created that can absorb and store the scant precipitation like a sponge.

Plants tolerant of shifting, abrasive sand take advantage of subsurface moisture. As a result, dune vegetation differs considerably from that of the surrounding playa or bajada. Shrubs and cacti are not common on unstabilized dunes because their slow-growing stems do not keep pace with the rate of burial by loose sand. Common perennial species instead tend to have creeping stems, rhizomes, or root stocks that maintain growing points just beneath the fluctuating

dune surface, protected from moving sand. These species include grasses such as Sonoran panic (*Panicum urvilleanum*) and big galleta (*Hilaria rigida*), shrubs such as sandpaper plant (*Petalonyx thurberi*), and the low-lying string plant (*Tiquilia plicata*). Ephemerals such as birdcage evening primrose (*Oenothera deltoides*), desert sand verbena (*Abronia villosa*), and purple-rooted forget-me-not (*Cryptantha micrantha*) are abundant in spring.

Some dune species are rare and unusual. A fleshy, leafless plant called sand food (*Ammobroma sonorae*) is a parasite, tapping into the roots and stems of several common dune shrubs to obtain a food supply. Three species at Eureka Dunes, in Inyo County, are found nowhere else on earth: the robust Eureka Valley dunegrass (*Swallenia alexandrae*), the delicate Eureka Dunes evening-primrose (*Oenothera avita* ssp. *eurekensis*), and the silvery Eureka milkvetch (*Astragalus lentiginosus* var. *micans*). Associated with these and other unique dune plants are some specialized and highly restricted sand-dwelling insects.

The presence of endemic plants and animals has led biologists to the conclusion that dunes are islands in a dry sea separated from the surrounding desert by unusual physical and biological conditions. These islands have formed and migrated during the climatic fluctuations of the last 20,000 years, exchanging dune-restricted organisms along short-lived archipelagos of sand blown between lowland basins. As populations on different archipelagos became isolated from one another, they diverged in form and function, selected by the local dune environment on which they were marooned.

These revelations about the natural history of desert dunes began to emerge in the 1970s, just as the number of off-road vehicles (ORVs) grew from a few thousand to 800,000 in California alone. Most ORV enthusiasts took their motorcycles and dune buggies to the desert, and sand dunes became a particularly popular destination. The once-silent sky was filled with the whining of engines as a profusion of tracks and trails spread onto the pristine landscape. The open, low vegetation of dunes, playas, and bajadas did not offer much resistance. Stems were broken and rootstocks torn from the soil by spinning, high-traction tires. Compaction of the desert soil prevented water infiltration and root penetration, and produced barren, highly eroded landscapes. Common and rare species of plants and animals were subjected to intense, mechanical disturbance, and entire biotic communities were dramatically degraded or destroyed over a short period of time. Ancient creosote bush rings, burrowing kit fox, ambling desert tortoises, and even sacred Indian rock drawings sustained the scars of thoughtless recreation.

In 1976 Congress passed the Federal Land Policy and Management Act, which directed the Bureau of Land Management to develop a management plan for thirteen millions acres of California desert. The California Desert Plan, signed into law in 1981, attempted to restrict off-road vehicles to certain dunes, playas, and bajadas in order to contain their impact and to preserve the special biological, geological, and archeological characteristics of these lands. It also addressed a wide

Unstabilized sand dunes support a vegetation that differs from that of the surrounding slopes and basins.

variety of other issues, including livestock grazing, mining, energy development, and wildlife protection. Although the plan was an important step in conserving the desert, it has been difficult to enforce because of a lack of personnel (only about fifty rangers patrol thirteen million acres).

Conservationists argue that the plan permits consumptive and degrading uses of the land that irreversibly impact vegetation and wildlife resources. Habitat destruction is the principal reason for the endangered status of thirty-eight kinds of desert plants and animals currently protected by state or federal law. An additional 296 desert species are candidates for protection or species of special concern because of their sensitivity to habitat destruction. More than a third of the desert communities recognized by the State of California are considered rare, sensitive to disturbance, and in need of special management consideration. Miners, ranchers, and ORV enthusiasts, however, fear that additional regulations would leave little room for economic and recreational development. Although patchwork amendments and small-scale legislation attempt to resolve these issues, we need to implement long-term, resource-centered management to eliminate the conflicting land use patterns that now drive species and communties towards extinction.

TRANSITIONAL VEGETATION

Traveling south through eastern San Bernardino County and into Riverside County, we begin to see species such as ocotillo (*Fouquieria splendens*) that characterize the hot Colorado desert. Many Mojave desert species are also present in this region, particularly creosote bush, Mojave yucca, burro bush, and big galleta grass. Unlike the Great Basin-Mojave transition, the blend of Mojave into Colorado desert is not marked by distinctive vegetation. Elements of both deserts usually are present, so the structure of the vegetation and its dominant life forms do not radically change. Although reasons for the subtle nature of this transition remain a mystery, some plant ecologists have suggested that gradients of summer rain and winter frost are responsible.

Hot Desert of the Lower Colorado Basin

The Colorado desert occupies low flatlands below 1,500 feet elevation. Summer heat is intense and often humid, and winter frosts are rare. It is only one part of the larger Sonoran desert, which extends from the Arizona-Utah border in the north to the Mexican state of Sonora in the south. The Colorado desert occupies the southeastern corner of California, and the Colorado River moves slowly through it, laden with red-brown silt, on its way toward the Gulf of California and the Sea of Cortez.

The Colorado River and its tributaries run to the sea, unlike the rivers of the Great Basin and Mojave deserts. This great river not only forms the political boundary between Arizona and California, it also separates two regions of the Sonoran desert: the Arizona uplands and California's Colorado desert. The Arizona desert, best observed around Tucson and Phoenix and into Sonora, is perhaps the most biologically rich and structurally complex desert on earth. It is far enough east to escape the immediate rainshadow of mountains in southern California and far enough south to experience subtropical storms. Annual precipitation (twelve to fourteen inches) is greater than in the Mojave desert and it comes in two seasonal peaks: a winter peak when temperatures are mild and a summer peak when temperatures are high.

Owing to the occurrence of both winter and summer rainfall, many unique plant growth forms can coexist throughout the year. Winter-deciduous and drought-deciduous trees, arboreal cacti, trees with green photosynthetic stems, rosette-shaped leaf succulents, evergreen and drought-deciduous shrubs, summer and winter annuals, and shrubby cacti all thrive here, each represented by a large number of species. These species combine to form an unusual, sometimes bizarre desert vegetation, with three canopy layers and a year-round pattern of biological activity.

West of the Colorado River, climatic conditions are different, and California's Colorado desert is only a poor cousin to the spectacular uplands of Arizona. Colorado desert vegetation does not possess the variety of plant growth forms, total number of species, or structural complexity of Arizona upland desert vegetation. The San Jacinto, Santa Rosa, and Laguna mountains form the western border of the Colorado desert and cast a rainshadow east across the Salton Sea and its low-elevation basin. Annual rainfall is reduced to as little two inches, and a smaller proportion falls during the summer (less than thirty percent, in contrast to fifty percent in the Arizona uplands). Midday temperatures exceed 100°F for weeks on end, and winter frosts are more frequent than in the Arizona desert.

Few plants can tolerate these extreme climatic conditions. Summer annuals depend on summer rain to carry them through a period of high temperatures. Other Arizona species, such as saguaro cactus (*Cereus giganteus*), are sensitive to low temperature and lack of summer rain. Some species even require a "nurse" for establishing their young: a rock, shrub, or tree that moderates seedling temperature by providing a shield against the hot summer sun and cold winter sky.

The broad transition between Colorado and Mojave deserts in California occurs along a line drawn between the towns of Banning to the west and Needles to the east. Near Essex, for example, the vegetation is dominated by creosote bush and burro bush, but an open and patchy overstory of smoketree (*Dalea spinosa*) and catclaw acacia (*Acacia greggii*) indicate departure from pure Mojave desert vegetation. A little farther south a greater variety of arboreal, shrubby, and succulent species is encountered. Ocotillo, California fan palm (*Washingtonia filifera*), jojoba (*Simmondsia chinensis*), ironwood (*Olneya tesota*), desert agave (*Agave deserti*), and chuparosa (*Beleperone californica*) are a few of the distinctive perennials of the Colorado desert not found in the Mojave. Although these tall species never form a closed overstory, Colorado desert vegetation often assumes a three-layered appearance. Low summer rainfall and freezing winter temperatures probably provide the climatic limits for Colorado desert species to the north and west.

SLOPE VEGETATION

Where vegetation of the Colorado desert is best developed, a tall (ten to thirty feet) but scattered overstory canopy emerges from an unusual assortment of tree species. It may be somewhat misleading to apply the word tree, which evokes an image of a solid, shady, and symmetrical growth form. Desert trees, such as smoke tree, have gnarled, open canopies with ragged edges and ephemeral leaves. Unlike the evergreen Joshua tree of the Mojave, arboreal species of the Colorado desert keep their leaves for only a portion of the year. Desert willow (*Chilopsis linearis*) is winter-deciduous and displays leaves only during the hot summer months. A long, woody taproot provides enough water for summer leaves by reaching deep supplies in or near washes. Species that tap permanent sources of ground water are called phreatophytes.

Another phreatophyte, palo verde (*Cercidium floridum, C. microphyllum*), is both winter- and drought-deciduous, producing a few small leaves in response to summer rains and shedding them as soon as soil moisture is depleted. Studies have shown that forty percent of annual photosynthesis by this plant is conducted in the green, chlorophyll-containing stems (hence the Spanish common name, which means green stick). Trees such as desert willow and palo verde seldom dominate the Colorado desert landscape, but they provide an intermittent canopy layer that is used by a wide variety of birds and small mammals.

Between and beneath the sparse trees of Colorado bajadas is an understory composed of tall shrubs, leaf succulents, and cacti. Creosote bush and burro bush dominate well drained slopes, along with a large number of other characteristic perennial species. Leaf-succulent agaves are often scattered below taller shrubs, such as jojoba (whose oily seed is commercially cultivated to replace whale oil in industrial applications), elephant tree (*Bursera microphylla*), ironwood, and ocotillo. Ocotillo is drought-deciduous, and it uses its green bark to photosynthesize in the absence of leaves. A new crop of leaves can be produced three to five days after a good rain. In winter and spring the leaves are followed by a profusion of bright red flowers pollinated by hummingbirds. As soil moisture is exhausted, withered leaves drop and ripe seeds disperse, leaving only barren, spiny stems to endure. A distinctive assortment of cacti is also found in the understory, including jumping cholla (*Opuntia bigelovii*), California barrel cactus (*Ferocactus acanthodes*), and pincushion cactus (*Mammillaria* spp.). On south-facing slopes cacti sometimes dominate the shrub overstory, producing a cactus scrub community.

A wide variety of plant growth forms is often associated with Colorado desert vegetation. The stem succulents seen here in Anza-Borrego State Park (San Diego Co.) include California barrel cactus (thick single stems) and jumping cholla (thin, branched stems). The tree-like ocotillo has drought-deciduous leaves, while shrubs such as creosote bush are evergreen. Agave species have succulent leaves and are often part of the understory (see page 177).

Above 1,500 feet in elevation the climate is cooler and moister. Bajada vegetation gradually becomes a mixture of shrub species from several geographic sources. This mixture includes some species from the Colorado desert—jojoba and Vasey sage (*Salvia vaseyi*), for example; some from the Mojave desert—desert senna (*Cassia armata*) and cheese bush (*Hymenoclea salsola*); and even some from chaparral—tree bladderpod (*Isomeris arborea*) and California buckwheat (*Eriogonum fasciculatum*).

A low herbaceous canopy containing many species is typical of Colorado slope vegetation. Perennial grasses such as big galleta, Sonoran panic, and Indian ricegrass contribute large amounts of cover where grazing has been kept to a minimum. Owing to the occurrence of summer rain, summer annuals can be diverse and abundant. One of these, chinchweed, is a small, spreading member of the sunflower family that posesses C_4 photosynthesis. It is able to grow when soil

temperatures exceed 100°F and to flower during August and September. Other summer annuals include sand spurge (*Chamaesyce ocellata* var. *arenicola*), Palmer's amaranth (*Amaranthus palmeri*), and magenta windmills (*Allionia incarnata*). Many winter ephemerals are shared with the Mojave, including bird-cage primrose, desert sand verbena, desert-sunflower, sand plantain (*Plantago insularis* var. *fastigiata*), Arizona lupine (*Lupinus arizonicus*), and desert-marigold.

Widely-spaced shrubs on this Colorado desert bajada surround open patches colonized by winter and summer annuals.

BASIN VEGETATION

Valley bottoms with saline soils have a simple basin vegetation composed of one or two canopy layers. If soils are coarse and the water table is deep, phreatophytic trees form an open overstory. Honey mesquite and and close relative, screwbean (*Prosopis pubescens*), are common. A monotonous understory of gray-green shrubs is often dominated by allscale (*Atriplex polycarpa*). If soils are fine-textured and the water table is shallow (perhaps six to nine feet from the surface), shrubs with succulent leaves endure the highly saline conditions. Iodine bush and inkweed form the only canopy layer of this simple, salt-stressed community.

Arroyo or wash vegetation is well developed in the Colorado desert. Desert willow, palo verde, honey mesquite, and catclaw acacia attain heights of fifteen feet or more and form woodland thickets where watercourses drain tall mountain ranges. Oasis stands of native California fan palm are found in the vicinity of the Salton Basin where non-saline soil water is ample. Large groves occur along major fault lines because breaks in the bedrock allow natural springs to surface. In contrast to almost all other desert species, palm leaves (fronds) cool themselves by transpiring large amounts of water. Frond temperatures can thus be well below air temperature and close to the optimum for photosynthesis at about 85°F. As old fronds die, they fold down and serve to insulate and protect the stem from intense sun. Lightning-caused ground fires remove the more flammable and heat-sensitive trees and shrubs from the understory and promote palm reproduction. Burning opens up the palm oases and permits growth of understory grasses (such as alkali saltgrass and alkali dropseed) and rushes. Proper management of California fan palm habitat should include the use of light, controlled burns.

A serious threat to Colorado arroyo vegetation is the invasive salt cedar (*Tamarix* spp.). Native to Europe and Asia, salt cedar colonizes washes, river banks, and playa margins. Introduced by Spanish explorers and settlers of the

seventeenth and eighteenth centuries, its rapid spread is aided by flash floods that disperse seed and disturb the soil surface. Along many miles of the lower Colorado River salt cedar forms an impenetrable thicket hundreds of yards wide. Salt cedar uses water at a high rate; a pound of leaves can transpire more than 1,200 pounds of water each year. Daily water use by a single tree is the same as a human family of four—and there is at least a million acres of salt cedar across the American west. Springs and subterranean water sources can be sucked dry by a developing thicket of salt cedar, thereby preventing the establishment of less competitive native plants. As the springs are lost, so are the bighorn sheep, pupfish, and other desert animals deprived of their water and native browse. Birds avoid the thickets, and yellow-billed cuckoo populations have been reduced from thousands to tens for lack of native riparian cover.

Arroyo vegetation near Mecca (Imperial Co.) is dominated by palo verde, ironwood, and cheesebush.

Although federal, state, and county agencies have recognized the salt cedar problem, eradication is difficult and costly. Removal is most successful if new invasions are stopped by digging up the saplings, roots and all, using a long-bladed trowel. Left alone, the established plants flower and produce large amounts of seed during their first year, making containment difficult. Large rootstocks quickly resprout if only the shoot is removed by cutting or burning. To prevent reestablishment of an even denser thicket after cutting, stumps

must be painted with an effective herbicide. This expensive, labor-intensive method is used only on a small scale, but it has been successful in Palm Springs, Death Valley, and Owens Valley. Other less costly methods of control will need to be devised to restore much larger riparian communities along the Colorado River.

Native people of the Washoe, Shoshone, Paiute, and Yuma cultural traditions inhabited desert land in California for at least 10,000 years. More than 12,000 native people used and depended on desert vegetation in the mid to late 1800s. Permanent settlements and transient campsites were constructed across the Great Basin and through the Mojave and Colorado deserts, using wood from desert trees: pinyon (*Pinus monophylla, P. edulis, P. quadrifolia*), palo verde, and mesquite. Over 350 species of plants native to the desert were harvested for food, from the common (basin sagebrush) to the rare (sand food). One group, the Cahuilla of southeastern California, utilized over 200 plant species for medicinal purposes alone. In these and other ways desert vegetation was the primary resource upon which subsistence and civilization were based. Despite the long duration, high intensity, and great importance of exploitation by native people, desert

The Changing Landscape

Groves of California fan palm (left) are restricted to canyons of the Salton Basin (Riverside, San Diego, and Imperial Co.) with ample natural springs.

Leaves of the California fan palm (above) cool themselves by transpiring large quantities of water.

vegetation was left essentially intact, unscarred, and productive. The disturbance created by these native societies in pursuit of economic and cultural growth was easily absorbed by the land and healed by natural processes.

European people have occupied desert land in California for less than 200 years. Permanent cities and ephemeral towns were built of materials imported from beyond the desert edge: yellow pine (*Pinus jeffreyi, P. ponderosa*), incense cedar (*Calocedrus decurrens*), and oak. Food, medicine, and other articles were

carried in on an expanding system of iron rails and graded highways. Native vegetation, no longer a resource of sustenance, was left to a diminishing population of native people and a burgeoning population of domesticated livestock. Aggressive plants were introduced from other continents. Water that once flowed from the mountains and through the desert was diverted, channeled, and exported, reducing some rivers to trickles and lakes to clouds of alkali dust. Mineral development, electric power generation, military training, suburb expansion, and motorized recreation became the principal human activities in California deserts. Vast areas of the modern desert, totaling millions of acres, were degraded or destroyed in a mere seven generations. The arid land no longer absorbs disturbance in the name of economic and cultural expansion, and many of the wounds will remain into and beyond the next century.

Can the desert heal itself after such severe human disturbance? Forest Shreve, an eminent desert ecologist of the early twentieth century, thought that natural restoration by community succession did not occur in deserts. Other ecologists have since found this to be false; succession does occur in deserts, but it is extremely slow. Studies of abandoned ghost towns have shown that short-lived species, such as cheese bush and rabbitbrush, take ten to twenty years to cover old roads and building sites. In some cases natural revegetation by short-lived species is sufficient to stop erosion and hide damaged landscapes. It does not, however, restore the structure or richness of the original desert community within the same amount of time. Long-lived species such as creosote bush and California barrel cactus require 100 years or more to assume their original abundance, especially if the soil has been compacted. Severe soil compaction by heavy vehicles may completely inhibit the recovery of desert vegetation, especially in areas that receive little rain. As a result, wagon train ruts from 150 years ago are still clearly visible. Scars left by General Patton's tanks during manuevers on nearly four million acres of California desert look fresh and barren nearly fifty years later. Denuded corridors follow hundreds of miles of pipeline and power transmission

Intensive use of the desert by off-road vehicles rapidly destroys the native vegetation. Recovery from such disturbance can be slow and incomplete.

lines. Natural restoration of mature desert vegetation is measured in centuries and cannot be regarded as a practical solution to human-induced disturbance.

Can the desert be artificially revegetated after disturbance? The answer is a definite maybe. Revegetation technology utilizes modern equipment and methods to rip compacted soil, drill-seed native species, transplant seedlings, and control weeds. One project required thirty years to reestablish six common species in a saltbush community destroyed by mining. Despite this intensive, long-term effort, only thirty percent of the target area had been revegetated with native species, and weeds such as Russian thistle, or tumbleweed, (*Salsola* spp.) became a serious problem.

Another project used ripping and drill-seeding to revegetate along the Los Angeles aqueduct in the western Mojave desert. After twenty-eight years artificial seeding had not consistently hastened or improved the natural recovery process. In some areas four-wing saltbrush, rabbitbrush, and even creosote bush became established after seeding. If young plants received heavy rain and protection from grazing and off-road vehicles, revegetation was even more successful. Other areas, however, showed almost no indication of having been seeded, and the vegetation was poorly developed or heavily infested with weeds. Artificial revegetation in deserts is expensive and requires a thirty- to fifty-year period of fiscal responsibility to ensure reasonable levels of success. Few private companies, utility districts, and govenmental agencies are willing to commit themselves to such long-term projects.

Clearly the desert will continue to be used for consumptive purposes that destroy or degrade its natural vegetation. Restoration is possible in some desert regions and vegetation types, but not in others. Proposals for development should be examined with this in mind and allowed to proceed only when all responsible parties can be legally committed to the successful restoration of diverse, self-perpetuating native communities. Otherwise, regions and communities for which we lack biological understanding or restoration technology should be conserved to prevent destruction by ignorance. Regions that support unusual assemblages of species should be conserved to maintain existing diversity at the community level. Regions that provide critical habitat for endangered plants and animals should be conserved according to law and our own sense of ethics and morality. Regions of the desert that tell us about the lives of native people should be preserved in honor of a heritage based on the bounty of a fragile land. Finally, regions of the desert that offer beauty and silence should be conserved for spiritual purposes and commodities of the soul.

7

NATIVE CALIFORNIANS AND CALIFORNIA VEGETATION

C ALIFORNIA'S VEGETATIONAL DIVERSITY IS PARALLELED BY THE cultural mosaic of its native people. For over 12,000 years waves of people migrated into California. Group after group dispersed along its coastline, upon its heartland prairie and inland waterways, and into its foothill woodlands. Other enclaves ascended its slopes and settled into alpine valleys, or descended onto its arid sagebrush basins and warm desert bajadas. These diverse settings created cultures of contrast. Early European travelers and scholars recorded evidence of more than 120 distinct and mutually unintelligible languages spoken.

The medley of native lifeways is an expression of cultural and sociological fitness between the people and their environment. Imagine, as writer Malcolm Margolin did, a typical summer's day in California several hundred years ago. On the great wind-riffled prairies Pomo and Wintun women stoop under the burden of pack baskets to dig roots from April's emerald landscape, or beat ripe seeds from the tawny grasses of August. In the oak-studded foothills, listen to the monotonous thump of stone pestles wielded by patient Maidu women grinding acorns. Down in the delta marshes and sloughs of San Francisco Bay a Yokuts woman uproots thick aquatic tubers while guiding her tule raft silently along many miles of curving, green waterways. High in the conifer-spired Sierra, an agile Miwok climbs a tall and stately sugar pine (*Pinus lambertiana*) to snap off the pendulous cones, while family members wait eagerly below to gather the pine nut delicacy. In the corrugated basin and range landscape of eastern California a small Washoe family cluster, weighted down with burden baskets, move on their annual trek up the Sierra's steep eastern escarpment to summer fishing and gathering encampments near thawing mountain lakes. Feel the heat of the Colorado Desert, as a Quechan family rests in the cool breeze and shade of their

California Indian tribes and territories, ca. 1700.

Miwok people made charcoal from the stems of virgin's bower (wakilwakilu) (right) to treat burns and sores.

ramada. Nearby, their maize, squash, and beans lie ripening on the warm and fertile mud of the Colorado River foodplain.

A Hunter-Gatherer Lifeway

While each of these Californian cultural groupings is unique, they merge into a collective whole by similarities in gathering and processing wild foods. Native Californians were hunter-gatherers. They subsisted primarily on wild plants and animals, and practiced agriculture only on a limited scale. Women usually foraged for plants and men engaged in fishing, hunting, and fowling. Men and women all over the state procured and prepared these wild resources in like fashion.

Many groups engaged in a pattern of seasonal mobility. Traveling on foot in small groups and with few possessions, they moved through different vegetation zones, their arrival at each coinciding with the ripening of particular plant resources. There was nothing casual or unplanned about their movements. Their gathering calendar was divided by season and was programmed with the flexibility needed to adjust to ever-changing seasonal and annual conditions. Although the quantity of wild plant foods was often sufficient to have supported many times the existing human population, wild foods could not permanently sustain a group in

Women of the Northern Paiute tribe display a variety of baskets in front of a bark-covered house. Photo taken in 1902.

a single locale for an entire year. Plants often grow in scattered locations, ripening quickly and simultaneously before many can be gathered. Yields vary from year to year, and their abundance is difficult to predict. A bumper crop one year is seldom followed by another.

Hunter-gatherers are not merely passive observers and collectors of the edible landscape. They are active participants in the landscape, practicing such methods as coppicing, soil and weed management, tillage, and burning. Shrubs such as redbud (*Cercis occidentalis*) and sandbar willow (*Salix hindsiana*) are coppiced to increase the number of slender branches useful in basketry. Rhizomes of certain ferns and sedges are known to grow long and straight in the absence of competing plants; these are removed by hand weeding. Historically, digging sticks were used to harvest bulbs and tubers, to loosen and aerate the soil, to propagate plants, and thus to improve subsequent harvests. Burning was used to manipulate shrub architecture, increase grass reproduction, improve access to harvest areas, keep vegetation open, and improve forage for animals hunted. In the past, burning was practiced in many grassland, marsh, scrub, and forest vegetation types. Such burning is not always possible today.

The food resources of California were bountiful in their variety. Although the acorn was the staff of life, free use was made of all kinds of roots, bulbs, corms, tubers, stems, leaves, and seeds. If one food supply failed, there were countless others to fall back upon. Native legends are nearly silent on famines. The Native Californian culture could be called the original affluent society. Yet these affluent hunter-gatherers were looked down upon by early travelers. Agriculture was considered the hallmark of civilization by Euroamericans, and nowhere in California did early explorers see the vast corn fields and multi-storied town complexes that DeSoto saw in Alabama or Coronado found in New Mexico. Rather, they observed native peoples who lived comfortably by harvesting what nature provided in her seasons.

Anthropologists have begun to investigate why agriculture was not extensively practiced in California, despite the fact that many groups displayed considerable sophistication in their knowledge of plant cultivation. Instead of viewing agriculture as the ultimate goal for all societies, researchers are now asking

Native American Florence Brocchini harvesting redbud shoots from the Sierra Nevada foothills during winter. The shrub was pruned a year before to make the long-straight, blood-red shoots she is now harvesting for basketry design.

A recently pruned sand bar willow along the Merced River.

the question: Why should hunter-gatherers become agriculturalists in the first place? In answer to this question, it has been suggested that California's beneficent environment actually inhibited the adoption of agriculture. Agriculture may have been an unnecessary alternative for Native Californians because of an already efficient economy based upon wild foods. In addition, these politically and economically complex societies had developed elaborate regional trading networks. Through trade, critical food resources that were not locally available were more widely distributed. Trade served to supplement food stores when local wild crops failed.

California supported the highest density native, non-agricultural population in North America. Although the state accounts for only one percent of the total area of North America, its rich environment supported ten to fifteen percent of the continent's population. (The modern population of California in relation to the rest of North America is about the same ratio today.)

Exploitation of human natives was as devastating as exploitation of the vegetatation. In 1769, when the first Spanish colonists arrived, we think there were 300,000 Native Californians living in the state west of the Sierran crest. Estimates by anthropologists range from 133,000 to nearly two million. We choose to follow estimates by Sherburne Cook (310,000) and D.H. Ubelaker (221,000), rounding up their average of 265,000 to an even 300,000.

Whatever the population in 1769, it dropped steeply for the next 150 years. The near extinction of native populations was consummated by the destruction of their native habitat by the fouling of streams, clearing of forests, draining of marshes, fencing of grassland, and replacement of natural plant cover by weedy introduced species.

Anthropologists hurried to salvage what they could from those who had witnessed the old ways of life. These early scholars were hindered by many obstacles. The native vegetation had already been reduced to a skeletonized version of the complex cover it once was. Much of California's vegetation had been changed by the invasion of alien species. Just as plant communities had been altered by foreign contact, traditional cultures were also transformed. Native survivors who disclosed ancient ways were recalling old customs second- and third-hand. Another barrier to an understanding of native lifeways was the selective reporting to which the accounts of early observers fell prey. The Victorian morals of some early writers inspired censorship of unfamiliar native ceremonials, sexual or clothing customs, and habits of cuisine, which often went unrecorded.

Finally, early plant lists were incomplete because predominantly male anthropologists tended to interview men. Traditionally, native women were the gatherers of plant foods, and the information they had was not usually accessible, or of interest, to men. The story of Ishi, the last Yahi, illustrates this fact. Ishi retreated from white civilization by hiding with his family in the foothills east of

Red Bluff. As the lone survivor of his family unit, he finally relinquished his native ways in 1911 when starvation forced him to surrender and join the dominant culture. He was cared for by anthropologists at the University of California in Berkeley. Ishi was of little help in the discovery of Yahi traditions regarding plant use, however. Perhaps if the last Yahi had been a woman, anthropologists would have had little trouble finding out about the intricacies of California native ethnobotany.

The story of plant and vegetation use that follows highlights only a fraction of the native plants of known importance to Native Californians. The scope of plants regarded as significant was wide and totals many hundreds. This vast inventory represents millenia of experimentation by trial and error. Furthermore, the Native Californians perceive each plant on a number of levels, according to the particular part of the plant used (root, stem, bark, leaves, seeds, fruit, pollen, and so on). A single species, valued as a plant food staple, might also be hewn into a shelter, fashioned into a garment, or crafted into a tool; it may also possess special healing powers or inspire an elaborate legend which acounts for the origin of the universe. This chapter focuses on food uses of plants in the native landscape.

Wetlands

In the aboriginal world—a world without metals, plastics, or synthetic fibers— native plants in varied forms served myriad functions and provided many different materials. Plants were integrated into every aspect of life: shelters, watercraft, clothing, dyes, weapons, cosmetics, and all manner of receptacles or containers.

The single most important tool in the technological repertoire of Native Californians was the basket. No other manufactured article touched their lives so completely, from cradle to grave. Marsh plants such as cattail or tule (*Scirpus acutus, S. californicus, S. validus*) were valued for their versatility in weaving everything from thatched houses to reed boats, sleeping mats, rugs, breechclouts, and diapers.

Native Californian babies were bound into a cradle-board basket carried by a broad band from their mother's brow. Children's toys were tiny models of the large baskets that they saw. Some native people lived in great thatched baskets— domed houses of willow and tule. They caught birds, game, and fish in woven nets and basket traps. Their seed meals were ground in bottomless mortar baskets, sorted in winnowing baskets, and screened in sieve baskets. Their nut soups were brought to a boil in large, water-tight cooking baskets. They ate from flat basket trays and drank from round basket bowls. When they journeyed, they packed their belongings in conical burden baskets. They navigated waters in large tule rafts. When they died, their bodies were draped in woven wares and priceless baskets were placed on their funeral pyres to accompany their souls to the other world. Baskets served the mundane tasks of everyday living, and they were also outlets for the highest expression of art.

A Maidu woman boils acorn mush in a water-tight cooking basket by plunging in hot stones. She holds the looped mush stirrer in her left hand.

With few exceptions, women the world over are the basket makers and weavers. Native Californian women are no exception. A woman who is especially skilled in basketry enjoys great prestige. The basketware itself reads like a book. Into the myriad weaves and shapes, the historical basket maker interwove her legends and beliefs. She copied nature in her designs. The surroundings in which the weaver lived and worked—features of the terrain, the varied plant cover, and resident wildlife—are expressed in the diversity of materials found in native basketry. The Cahuilla employed yucca fibers and fine grass; the Paiutes, coarse tule and cattail; the Pomo, sedge rhizomes and willow stems; the Yurok, bracken fern (*Pteridium aquilinum* var. *pubescens*); and the Maidu, redbud and Jeffrey pine (*Pinus jeffreyi*) roots. Historic methods are still used by modern basket makers.

Collecting and preparing plant materials for a basket are as involved as the weaving. Sedge root beds (really rhizome beds) are cultivated by removing rocks, other roots, and debris, loosening the soil, and transplanting root sections. Quality stands of utilitarian plants such as willow are tended by pruning and thinning and harvested year after year.

Baskets—both coiled and twined—are commonly woven on a framework of slender willow shoots which are peeled and carefully cured. Other plants provide color to the design. Black patterns are created by the long running rhizome of bracken fern. A rich red-brown design element is obtained from the bark of redbud. Roots of foothill pine (*Pinus sabiniana*) or Jeffrey pine contribute a warm, creamy tint. The dark orange rootstock of Joshua tree (*Yucca brevifolia*) also finds a place in coiled ware patterns.

Historically, marsh plants also filled a variety of nutritional needs. The starchy rootstock of yellow pond lily (*Nuphar polysepalum*) was an important food source for people living in northeastern California. The large floating leaves and huge yellow flowers once covered thousands of acres of shallow lakes, marshes, and slow-moving streams in the territory of the Modoc, Achomawi, and Atsugewi. The tubers were harvested from the great Klamath marsh by canoe. Modoc women, wading in the water and pushing small canoes before them, loosened the tubers with their toes so that they floated to the surface. Roots baked in the embers of a fire were skinned and eaten whole or mashed. Women also plucked seed pods from the flower centers, to be roasted and ground into flour.

For coastal groups, marine plants were an important source of vitamin- and mineral-rich greens. The Yurok gathered algae, pressed it into cakes when wet, and ate it dried. Seaweed fronds were either baked or cut and pressed into an uncooked mass to be nibbled with meals. Seaweed is tough, and chewing it is a tedious process, but it provides a vital source of salt and iodine.

Succulent perennial plants of coastal salt marshes and wet tidal flats are filled with brine. The fleshy stems of pickleweed made a pleasant, salty salad green for Coast Miwok and Costanoan peoples. Saltgrass (*Distichlis spicata*) was, and still is, valued for its salt content. The dew on its blades is rich in salts excreted from the leaves and evaporated into crystals. Salt crystals can be collected by threshing the blades; they provide a gray-green seasoning with the flavor of a salty dill pickle.

Grasslands

Perennial bunchgrasses have many uses. The stems of some are prized for basketry, while others have been used for the thatch of an acorn granary, mats, and bedding. The nutritious seeds of native bunchgrasses were once of foremost importance in the aboriginal diet. Large quantities could be obtained and stored for long periods of time. Chia (*Salvia columbariae*) seeds carbon-dated to over 600 years old have been found in pottery jars at burial sites on Santa Rosa Island.

In traditional seed harvest, women used woven seed beaters shaped in the form of a short-handled tennis racket to beat bushels of ripe seeds into large, conical burden baskets. When the stalks are ripe, they shatter easily, and women moved the baskets through the grass in order to collect the seeds. Threshing was necessary to separate seeds from unwanted chaff, so the day's collection had to be vigorously churned. Churning caused the lightweight chaff to accumulate on the surface where it could be easily removed. The coarse hulls were then singed off. This threshing process was repeated until the grain was largely clean. The seeds were then parched by skillfully tossing them about with live coals in a shallow

Purple needlegrass.

basket tray. Parched seed was ground into a fine flour on a flat rock milling slab with a handstone. Different kinds of seeds were often blended to suit the most fastidious taste.

Grasslands which produced this bountiful seed crop were managed by aboriginal Californians who deliberately and regularly set them on fire. The Chumash, and many other California groups, understood the connection between fire and the growth of certain useful plants. Regular burning maintained the dominance of bunchgrass in environments where chaparral or coastal scrub vegetation would otherwise have prevailed. Fire promoted seed production by removing dead thatch and enriching the soil. It encouraged vigorous growth of seed plants, bulbs, greens, and shoots as well as the production of long stems or flower stalks useful in basket making. Deergrass (*Muhlenbergia rigens*), the most important bunchgrass for coiled baskets, historically was managed by burning. Fires were set in late summer, after seed stalks had been collected, and the same area was repeatedly burned every three to five years.

The journals of early explorers and missionaries provide information useful in our reconstruction of aboriginal management of the wild California landscape. Father Juan Crespi accompanied the Portola expedition in 1769-70, and he made detailed written observations of native cultures and natural environments. In his journal entry of August 20, 1769, he wrote:

> We went overland that was all of it level, dark and friable, well covered with fine grasses, and very large clumps of very tall, broad grass, burnt in some spots and not in others; the unburned grass was so tall that it topped us on horseback by a yard.

Other descriptions present a picture of a park-like landscape with scattered oak trees that was sustained by deliberate and frequent burning by Native Californians. Father Crespi noted in May 1770 that inland from the Santa Barbara coast, "not a bush was seen." Today this area is covered by coastal sage scrub and chaparral. The shift from grassland to chaparral resulted from fire suppression by early settlers. A proclamation by Governor Jose Joaquin de Arrillaga in May 1793 addressed:

> the wide-spread damage which results to the public from the burning of the fields, customary up to now among both Christian and Gentile Indians in this country, whose childishness has been unduly tolerated. . . I see myself required to have the foresight to prohibit for the future. . . all kinds of burning. . .

Chia, a member of the mint family, produces seeds that were an important cereal for native Californians.

Native bunchgrasses that may have been most important before the introduction of weedy, alien species include California brome (*Bromus carinatus*), meadow barley (*Hordeum brachyantherum*), and needle grass. Rice grass (*Oryzopsis hymenoides*) was a staple grain for the Washoe and Paiute in eastern California.

When boiled, these grass seeds swelled like rice. They were also toasted, pounded into a flour, and cooked into mush. Pacific beach grass (*Elymus mollis*) was important along the coast—from Monterey to Mendocino—to Yurok, Salinan, Costanoan, and Coast Pomo peoples.

Seeds of several members of the sunflower, amaranth, goosefoot, and mustard families and chia, of the mint family, also were ground into flour and made into mush or bread. These seeds are as rich as butter, and the oily seed meal could be picked up with the fingers in lumps.

It is because Native Californians dug bulbs, tubers, and corms in such profusion that they were once disparagingly referred to as digger Indians. The plants that produced these underground organs (also called Indian potatoes) were common associates of bunchgrasses in coastal prairie, Central Valley prairie, and foothill woodlands. In spring, dances were called to celebrate the coming harvest. Once these first-food ceremonies were completed, women armed with fire-hardened digging sticks went into the wet meadows to dig up the bulbs. A straight and pointed stick, preferably made of mountain-mahogany or other extremely hard wood, was charred in the fire and rubbed to a sharp point on a stone. The most abundant and widespread bulbs were of brodiaea (*Brodiaea* and *Dichelostemma* spp.), mariposa lily (*Calochortus* sp.), and common camas (*Camassia quamash*). Pristine stands were so profuse that one early writer counted over 200 bulb plants within one square foot of ground. Native women regularly left the smaller bulbs, tubers, and corms in the field, harvesting only the larger ones. The small ones grew and flourished, maintaining the area for the next year.

In spring bulb plants can be easily spotted by their brilliant blooms. Both common camas and death camas (*Zigadenus venenosus*) grow in the same area, and their bulbs and leaves look alike, but flowers of common camas are blue, while those of death camas are cream-colored.

The deliciously nutty camas bulbs are occasionally eaten raw, but historically they were generally steamed first in earth ovens. To bake these delicacies, a pit was excavated with a digging stick. The pit, lined with stones, held a fire built inside. More hot stones, heated in a fire built beside the pit, were placed on top of the coals, and the cleaned bulbs then placed on a mat of fresh maple or mule ear leaves. Another mat of madrone (*Arbutus menziesii*) leaves, soap plant (*Chlorogalum pomeridianum*) leaves, or branches of poison oak was laid on top. More hot stones were added, and the entire pit was covered with earth. Water poured around the edges worked its way down to the hot stones and produced steam. The pit was opened the following day and revealed a sweet feast. Women pressed the molasses-like cooked mass between their palms into dessert cakes that looked like macaroons or ginger snaps. The vanilla-like fragrance and maple sugar flavor of the bulbs were relished; they were dried for later consumption on special occasions.

Wild onions (*Allium* spp.) were and still are harvested by Native Californians

Bulbs, Tubers, and Corms

and non-native backpackers alike. Growing in thick patches in montane meadows and on dry, rocky plains, they can be easily recognized by their characteristic onion odor. Both the bulbs and the green stems have a strong taste. The bulbs are not large and round, as are modern cultivars available in food stores, but take the form of an elongated, thickened rootstock no larger than a child's thumb. They grow more shallowly than brodiaea bulbs and are more easily harvested.

Yampah (*Perideridia* spp.) has finger-like tuberous roots with a mild nutty flavor and can be eaten raw. Modoc and Pit River peoples of northeastern California relied heavily on this herbaceous perennial for carbohydrates. The tuberous roots were peeled, boiled, and eaten like potatoes. The dried tuber was also ground into flour.

Oak Woodlands

Oak woodlands were a source of basketry material, arrow wood, tools, construction and fish trap material, household utensils, and—of course—food. Acorns were the primary staple of Native Californians, just as rice and corn are the staff of life for Asians and Mesoamericans. In the past, acorn eating was the hallmark of groups such as the Maidu, Miwok, Wintu, and Pomo who inhabited the foothills of the Sierra, Transverse, Peninsular, and Coast ranges. There are about twenty species of oaks throughout California, but only nine were of economic importance to Native Californians: tanbark oak (*Lithocarpus densiflorus*), valley oak (*Quercus lobata*), Oregon oak (*Quercus garryana*), blue oak (*Quercus douglasii*), scrub oak (*Quercus dumosa*), canyon oak (*Quercus chrysolepis*), coast live oak (*Quercus agrifolia*), interior live oak (*Quercus wislizenii* var. *wislizenii*), and especially California black oak (*Quercus kelloggii*). All acorns were not valued equally, so personal preference and relative abundance dictated which species were the principal ones harvested in any given area.

In fall acorns were knocked from oaks with a long pole, collected from beneath the trees, and thrown into large, conical baskets. These burden baskets rested on the back but were suspended from the forehead by a broad band, which freed both hands to gather acorns and toss them backwards into the basket. Proficient harvesters could gather up to seventy-five pounds of acorns in an hour.

Tiapi-Ipai woman near Cuyamaca (San Diego Co.) grinding with mano on metate. A basket tray catches any fallen meal. Photo taken ca. 1905.

Historically, acorns were either dried in their shells and stored whole in acorn granaries or they were hulled and ground. Women congregated at the communal milling area to grind and to chat. The brittle shells were first cracked open on an anvil stone. Nut meats were removed and pulverized by pounding a heavy, cylindrical stone pestle up and down into a portable stone mortar bowl or into a cup-shaped depression worn in a bedrock outcrop. Often a bottomless basket, held in place with the legs, was positioned over the shallow cup-shaped depression to prevent particles of acorn meal from flying away. Coarser particles were removed from the fines for regrinding by sifting them in a large, shallow basket. Any stray bits of meal that dropped around the basket were swept up by a brush made of soap plant

fibers. New technologies, such as coffee grinders, simplify this process today.

The most time-consuming task in the preparation of edible acorn meal is the lengthy process which removes objectionable and poisonous tannic acid. Ground meal is bitter, puckery, and unfit to eat until tannic acid is removed by leaching with water. To do this in historical times, a hollow of clean, white sand was scooped out. Acorn flour was poured in and covered with cold water. Branches of

Acorns of coast live oak (top), and canyon oak (middle).

Hupa woman leaching acorn meal in a basin of sand. She holds a basket so tightly twined that it holds water. Photo taken ca. 1902.

incense cedar were positioned to channel water evenly and gently over the meal; cedar also provided a balsam-like flavor. In two to three hours the water percolated through the meal and sand, carrying with it a portion of the bitterness. This percolation process—much like making drip coffee—was repeated until the flour was rendered perfectly sweet. The leached meal was removed by placing a hand on top of it, palm downward and fingers spread to fullest extent. The flour adhered to the hand, which was then washed off into a large cooking basket. Modern-day Native Californians run faucet water through a filter bag to leach acorn meal.

Miner's lettuce, an important salad green.

In the past acorn meal was placed in water-tight baskets set into a shallow pit to prevent capsizing. Water was brought to a boil by repeatedly plunging in dozens of hot cooking stones, which were stirred constantly so as not to burn the basket. Acorn soup could be eaten plain or flavored with berries, nuts, meat, insects, ash, or clay. Depending upon its consistency, soup was consumed by dipping one, two, or three fingers into a communal mush basket.

Bread made from acorn flour is deliciously sweet, rich, and oily. The prolonged and gentle baking converts some of the carbohydrates into sugar. When removed from an earth oven, it is a jet black throughout. While still fresh it has the consistency of soft cheese, but in the course of a few days it can become quite hard. Bearing the imprint of its leaf wrappings, a black lump of bread can be easily mistaken for a fossil-bearing piece of coal.

The food value of acorns is quite substantial (see table, p.183). Acorns are slightly lower in carbohydrates and protein than grain, but they are richer in fat, fiber, and calories. The fat content far exceeds that of almonds, and the caloric content is greater than that of chestnuts, filberts, coconuts, and Brazil nuts. Acorns are also high in calcium, magnesium, and phosphorus. John Muir was sustained by hard, dry acorn bread during his treks in the mountains around Yosemite Valley, and he deemed it the most strength-giving food he ever ate. The robust physique of most acorn-eating native peoples testified to the value of acorns as fatteners. The quantity of acorn meal cooked and eaten astonished early Euroamericans in California. Each family could consume 500 pounds or more per year.

Greens

Plant species eaten as greens by Native Californians far outnumber those used in modern diets. They occur in many plant communities, from sea level dune swales to high mountain meadows. Our examples here come from the foothill woodlands. In the past most were consumed after stone-boiling in a basket or after steaming in an earth oven, although some were eaten raw.

Among the first wild greens to appear in spring are fiddleheads, the young shoots of bracken fern. Bracken is the best known and most common California fern. The tender, young, unrolled fronds, picked when still under six inches tall, were boiled or steamed until tender.

Fruits of Christmas berry (left), elderberry (below), and Manzanita (bottom).

Historically, clover (*Trifolium* spp.) was eaten in great quantities raw, steamed, or boiled. From early April to July entire village populations ranged through clover-rich meadows, gathering it by the handful. After enduring a long winter on a short supply of greens, they filled their stomachs with fresh foliage and crisp stems. Bloating often resulted, and it was treated by a mixture of soap plant extract and salt water.

Miner's lettuce (*Claytonia perfoliata*) is still a popular salad green of traditional Native Californians. The leaves and stems are tender and succulent and taste much like spinach when cooked. It was adopted from the native people by gold rush immigrants who used it to stave off scurvy.

Many wildflowers also served as greens to aboriginal Californians. The young leaves and stems of monkeyflower (*Mimulus* sp. and *Diplacus aurantiacus*), columbine (*Aquilegia* sp.), paintbrush (*Castilleja* sp.), California poppy (*Eschscholzia californica*), and lupine (*Lupinus* sp.) were eaten as a salad or cooked as greens.

Berries

Wild berries abound in the foothills of California, in northern coastal scrub, chaparral, and oak woodland communities. Sweet berries can be eaten raw or dried into cakes and stored for later use. They were pounded into a flour, cooked into a sauce or syrup, or made into a soft drink by gentle bruising and steeping in

water. Many of these wild berries are still appreciated and collected today: blackberries (*Rubus* sp.), western raspberries (*Rubus leucodermis*), thimbleberries (*Rubus parviflorus*), salmon berries (*Rubus spectabilis*), gooseberries and currants (*Ribes* sp.), and elderberries (*Sambucus* sp.).

Christmas berry (*Heteromeles arbutifolia*), also called California holly or toyon, is a common chaparral and oak woodland shrub. The now famous city of Hollywood was named for the large number of California hollies which grew on the surrounding hills. The somewhat bitter red berry clusters are rarely eaten raw; they must be roasted or parched to remove the bitterness. Early Spanish settlers also cooked these fruits by boiling and steaming.

More than forty species of manzanita (*Arctostaphylos* spp.) grow together with toyon in chaparral and oak woodland communities, and most of their berries can also be used as food. Manzanita means little apple in Spanish, a name that refers to the berries that are no more than one-third of an inch in diameter. Traditionally, their ripening is heralded by the Maidu with special dances and a "big eat." The little fruit is dry and mealy, but nutritious. It is ground into a powder and made into bread or mush. Manzanita cider is made by first pounding ripe berries into a flour. The seeds and skins are removed, and the flour soaked in water for a long time. The flour is then heaped up on a basket into a mound, a crater formed in the center, and a minute stream of water poured into the crater. The water, which slowly percolates through the meal, is transformed into a delicious beverage.

Bulbs, Tubers, and Corms

Most of the bulb plants harvested in low-elevation grasslands also grow in woodlands. A few, such as soap plant, are more abundant in woodland than in grassland, but few herbs are completely restricted to woodland.

Soap plant is unique in the diversity of uses its bulb provides. Its large bulb is covered with dark brown, hairy fibers. The inner portions of the fresh bulb are slippery, and when crushed and rubbed with water, they froth like ordinary detergent. The absence of any alkali in soap plant roots makes it preferable for washing delicate items and baskets. Used as a shampoo, it controls dandruff and leaves hair soft and glossy.

Historically, the fibers around the soap plant bulb were bound together into a coarse whisk broom for sweeping up acorn meal around grinding holes. The brush was glued together with the sticky extract of the plant. This gummy substance was also applied externally to relieve cramps, rheumatism, stomach ache, and to heal sores; it was taken internally as a diuretic and laxative.

The soap plant bulb is a good source of carbohydrates and can be eaten if roasted for more than thirty-six hours. Uncooked, bulbs contain a nerve toxin useful for catching fish. Large numbers of fish can be obtained with little effort by crushing raw bulbs which release a lather in dammed streams. Fish become stupefied, float to the surface, and can be eaten without ill effect.

Soap plant brushes.

Other important woodland bulb plants include brodiaeas, wild onion, mariposa lily, and yampah. Widely distributed, they are discussed in grassland and forest sections of this chapter.

Montane Forests

One does not usually think of conifer trees and forests as sources of food. To Native Californians of the past, however, these forests were great orchards. While oak acorns symbolically provided bread for subsistence, pine nuts provided the cake.

Many species of California pine trees were and are important in a variety of cultural ways. Pine nuts (seeds) can be devoured raw, roasted, boiled, or pounded into a rich, dark, oily butter. Historically, pine nuts were a major article of commerce and trade. The hard shells were perforated and made into ornamental beads which decorate Hupa, Karok, and Atsugewi dance regalia. The cambium, or inner bark, was stripped and eaten in times of food shortage. Pine pitch, which oozed from bark punctured the previous season, was a valuable adhesive, sealant, and antiseptic. It can also be boiled into a tea to cure colds, rheumatism, tuberculosis, flu, indigestion, fever, nausea, kidney and bowel disorders, and can be used as an eye wash.

Sugar pine sap is rich in sugar and can be used as a chewing gum. Its pale brown feeder roots still are woven into Pomo baskets. Fallen trunks fashioned Yurok water craft or Miwok houses. Early Euroamerican pioneers, including the renowned Donner Party, could have collected and eaten pine nuts along much of their route across the mountains, but their food prejudices and ignorance caused needless starvation.

Several species of pine were most heavily used by Native Californians in the past. Among these are the pinyons (*Pinus monophylla, P. edulis, P. quadrifolia*)—a group of short pines that grow where mountain slopes meet the desert's edge. Though small in size, pinyon pines produced the food, medicine, and building materials enabling native cultures of eastern California to flourish. A single Paiute family might harvest 300 pounds or more of pinyon nuts for winter use. The Paiute and Washoe of northeastern California celebrated the pinyon harvest with festivals of feasting, dancing, gambling, trading, and courting. John Muir described a Mono Lake Paiute pine nut harvest in the 1870s:

> When the crop is ripe, the Indians made ready the long beating-poles; bags, baskets, mats, and sacks are collected; . . . old and young, all are mounted on ponies and start in great glee to the nut-lands. . . Arriving at some well-known central point. . . the squaws with baskets, the men with poles ascend the ridges to the laden trees, followed by the children. Then beating begins right merrily, the burrs fly in every direction, rolling down the slopes,

Soap plant.

lodging here and there against rocks and sagebushes, cached and gathered by the women and children with fine natural gladness. Smoke-columns speedily mark the joyful scene of their labors as the roasting fires are kindled. . .

On the western side of the mountains foothill or ghost pine was harvested by the Shasta, Yuki, and Yana tribes. Like pinyon cones, foothill cones are tightly attached to branches and must be knocked off with a stick or twisted off by hand. The mature cones can be collected in October, stacked in piles with tip ends down, and covered with pine needles. Traditionally, the piles were ignited to remove the pitch and make the cones easier to handle. Seared cones were then split open and the seeds beaten out. Seeds, roasted or baked in an earthen oven, can be dried and stored for later consumption. In June green cones can be pulled down and the seeds eaten fresh. The pithy center of the green cone, roasted in hot ashes, yields a brown, viscous, slightly syrupy food.

Pinyon pine branch with cones.

While pinyon and foothill pine nuts served as a daily staple for many tribes, the seeds of lofty sugar pines were regarded as a delicacy. Agile Miwok climbers ascended sugar pines with the help of a hooked stick, or by placing a small dead tree against the trunk to serve as a ladder. Sugar pine cones are the longest in the world and they hang pendulously in clusters at the ends of horizontal limbs. They are of sufficient weight that climbers could remove them by setting the branch in a rotating motion with one foot until the cones snapped off.

Other species of pine that have large seeds and are important food sources include Torrey pine (*Pinus torreyana*), a coastal species with a narrow distribution in southern California, and Coulter pine, a dominant tree in the mixed evergreen forest of the Coast, Transverse, and Peninsular ranges. Higher montane species, such as ponderosa pine (*Pinus ponderosa*) and Jeffrey pine, are also part of the harvest, but they have smaller seeds and are not as centrally important in the menu of seed resources.

Pine seeds have outstanding food value, and are especially high in fat content (see table, p. 183). Modern diets tend to focus on protein and to play down fat, but fat is highly important to people exposed to low temperatures and lacking warm clothing, as was the case among aboriginal Californians. The fifty percent fat content may have been a more important survival factor than the twenty-five percent protein content.

Desert Scrub

Native Californians used desert plants in as many ways as tribes to the west used foothill woodland plants. One of these important uses was as food. Arid desert bajadas and basins appear to produce little food. However, desert plants sequester nutrients and moisture in reserve, which can be harvested by the cunning gatherer.

The Mojave, Yuma, and Cahuilla traditionally ate mesquite as a staple legume. Honey mesquite (*Prosopis juliflora* var. *torreyana*) and screwbean mesquite (*Prosopis pubescens*) can drop up to 100 pounds of straw-colored pods under one tree in a good year. These pods could be dried in great quantities and packed away in basket granaries. Pod and seed were pounded together into a coarse meal which contained up to thirty percent sugar. The flour was soaked in water, and the resulting light fermentation improved the flavor. Rolled into compact balls, this nutritious food could be handily packed for a journey. Eating mesquite balls must have required vigorous chewing. Traditionally, a fermented beverage was made in a wide clay basin, kept filled with water and half-crushed pods. Everyone helped themselves to the liquid during hot summer months.

Agave (*Agave utahensis, A. deserti, A. shawii*)—also known as mescal or century plant—is another desert food staple. Several species grow in the Colorado and Sonoran deserts of California, Arizona, and Mexico, where they are made into the alcoholic beverages pulque, vino mescal, and tequila. For a decade or more agaves grow vegetatively, producing large, stiff leaves armed with a spine at the tip. Then the plants produce a flowering stalk twenty or more feet tall. At first the stalk

Agave, a spiny leaf succulent of the Colorado Desert used for food.

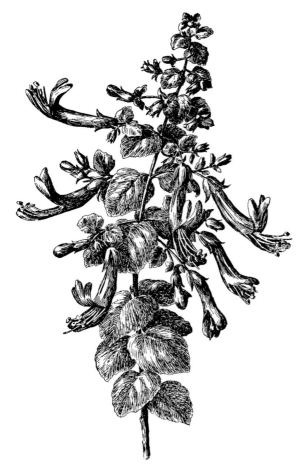

Climbing Penstemon.

resembles a giant asparagus shoot, but then the flower buds open in a burst of yellow at the apex. After the flowers have pollinated and set seed, the plant dies. The flower stalks are filled with carbohydrate-rich sap. Young stalks are roasted in great earth ovens dug into sand and lined with hot stones. The stalks, covered with grass and earth, are left to roast for a day or two. When cooked, the fibrous tissue is transformed into a sweet, molasses-colored material that can be stored for years. The blossoms also can be collected, boiled, dried, and stored for later consumption.

Cacti such as beavertail (*Opuntia basilaris*) and prickly pear (*Opuntia occidentalis*) are prized for their fruits. In early summer the fruits are broken off with a stick and their fine, short spines are brushed off with a bunch of grass or twigs. Fruits can be cooked or steamed in an earthen oven for twelve hours or more. The flesh of the cactus pad is a source of liquid. It can also be pounded into a wet medicinal dressing.

The only Native Californians who engaged extensively in agriculture were desert tribes living in Owens Valley, the Salton Basin, and the Colorado River delta. They were part-time farmers who used their produce to augment a hunter-gatherer lifeway. The Kamia of the Imperial Valley and the Cahuilla of the Coachella Valley employed ditch irrigation to capture floodwaters of the Colorado River and the small streams which descended desert-facing slopes of the Peninsular Range. Crops included nutsedges, maize, beans, and squash.

Maladies and Medicines of Native Californians

Medicinal plants come from every plant community. The patient quest for food and the minute inspection of every growing thing to determine any possible useful properties led to the discovery of medicinal herbs. Native Californians were fully aware of the lethal properties of certain plants, and they removed the poisons of some to render them edible. Administering medicinal plants with poisonous properties usually was done by a trained specialist who practiced great caution and moderation in the prescription.

If a disease was severe, chronic, or exotic, the afflicted person or his or her family solicited the services of a medical practitioner or shaman. Cures were invested with ritualistic behavior designed to connect with some supernatural or special agent, and they were aided by medicinal plants. The aromatic leaves of basin sagebrush, for example, were valued as a talisman to ward off evil spirits and ghosts.

For more common maladies—wounds, skin irritations, indigestion, tooth-aches, snake bites, eye inflammations—treatment had a more empirical basis, and plant remedies tested by generations of experimentation were employed. Some native drug remedies recently have been analyzed for the pharmaceutical properties of their oils, tannins, acids, alkaloids, and other chemical components. These analyses have substantiated the effectiveness of native home remedies. For example, a decoction made from the roots of Oregon-grape (*Berberis aquifo-lium*)—a shrub in the north coast forest—was commonly used by Native Californians as an eye wash. Chemical analysis and present-day usage both suggest that its application is effective in reducing inflammation of the mucous membrane of the eye. Some commerical eyewashes today contain the same active ingredient.

Native Californians continued to rely on native medicinal plants long after contact with Euroamericans forced them to change many other aspects of their lives. This is documented by ethnobotanical plant lists, collected throughout historic time, which show a steady decline in the total number of native plants used for food, while medicinal plants contin-ued to be well represented.

Unfortunately, traditional therapeutic practices were little match for the onslaught of new illnesses such as tuberculosis, smallpox, whooping cough, and venereal disease. Health problems were further aggravated by wretched living conditions and dietary deficiencies among displaced peoples. As new diseases were contracted, the medicinal plants that had been successful cures for traditional ailments were tried with little effect. The situation is tragically illustrated by the large number of plants unsuccessfully employed to cure syphilis and gonorrhea: paintbrush (*Castilleja* spp.), Indian balsam (*Lomatium dissectum* var. *multifidum*), phlox, angelica, penstemon, juni-per, Mormon tea, quaking aspen (*Populus tremuloides*), black cottonwood (*Populus trichocarpa* var. *trichocarpa*), and others. Leaves of black cottonwood also formed the essential ingredient in a traditional love potion. An old

Yerba buena, "good herb," was an important medicinal herb for Indians.

Shoshone herbalist summarized the problem: ". . . in the past, one plant was used for one disease. Now, there are too many diseases since Columbus came. So we try anything for everything."

Ceremonial Plants

Native Americans engaged in drug use cautiously, and only after rigorous observance of preliminary sex and food taboos. The aboriginal drug user usually was accompanied through the altered experience by a constant companion or guide. For Native Californians, the taking of drugs was integrated into a backlog of cultural—not recreational—tradition. The drug served as a large part of the training to cultivate spiritual emotions, to understand the mystery of life, and to appreciate the reality of the unseen. It was an enhancer of life, not an escape from it.

Jimson Weed

One drug of multiple uses was datura (*Datura meteloides*), or jimson weed, a toxic plant with hallucinogenic and narcotic properties. Several species occur in different parts of California. Datura grows over much of southern California and the lower Sacramento Valley, and is a component of grassland vegetation. Its flaring, trumpet-like, white flowers are a conspicuous feature in disturbed parts of the grassland. Jimson weed was taken as a medicine for serious injuries and lingering illnesses. It was an anesthetic and a charm for broken bones and wounds, and it was made into a poultice and a blood tonic.

Jimson weed was also taken to induce delirium and produce a vision which established contact with a supernatural guardian or dream helper to obtain good luck, to communicate with spirits of the dead, to foretell the future, and to avert misfortune foretold by some ill omen. Youths among the Gabrielino, Luiseno, and other south coastal groups took it as part of a puberty ritual, one which was integrated into a much broader ceremonial complex called the Chingichnich cult. Farther north, among the southern Yokuts and western Mono, men and women drank datura as part of a spring ritual. The Mojave, Yuma, and Cahuilla had no well developed datura ceremony and took it in any season. It is possible that some of the rock paintings found at remote sites in Chumash territory were created under the influence of datura. Their brilliant colors and fantastic figures might well depict datura-induced visions.

Jimson weed is dangerous, and it was carefully procured and prepared. The margin between a narcotic and a fatal dose is narrow. Chumash who dug the root purified themselves first and performed the proper incantations before carefully unearthing it. To prepare the drink, a slightly roasted root was mashed in a special ceremonial mortar and steeped in cold water. It was quickly drunk from a basket without taking a breath; otherwise its smell would discourage drinking enough of it. The datura-giver had to carefully calculate the dose according to the soil in which the plant grew, the plant's age and size, the concentration of the brew, the season of the year, and the body size of the drinker. The Yokuts dug the root only

in winter and early spring; they thought it too strong later in the year. Deaths from overdose were not unknown, despite long familiarity with the plant. Responsibility for a death fell on the drinker, not the giver. The Chumash believed that a datura drinker who died had violated one of the taboos and aggravated the datura spirit, or else he "lost the trail" and did not return from the spirit world.

Tobacco and datura were among the first medicines of Native Americans. Tobacco was in wide use in the Americas when the Spanish arrived. It was later taken to Europe, Africa, and Asia, where it became an item of pleasure in contrast to its religious and social significance for aboriginal Americans.

Tobacco

At least two of the six California species of tobacco (*Nicotiana* spp., especially *N. bigelovii* and *N. attenuata*) were used throughout the state. Native tobacco has a harsh, strong flavor. Native Californians were light smokers compared to native peoples of the eastern United States. Small quantities of tobacco were smoked by shamans during their rituals to aid supernatural activities. Tobacco could beguile one's enemies or bring one luck in gambling or in hunting. Men found a smoke

Datura or Jimson weed.

in the sweathouse to be euphoric and beneficial to sleep. Smoking served as a greeting of friendship along the trail. Tobacco was burned in the campfire to keep unfriendly spirits at bay. It was also used as a standard of exchange.

Dried leaves were pulverized, powdered into snuff, or moistened and compressed into a solid lump. Small bits were then broken off and eaten by men, women, and children. A drink made with tobacco served as an emetic. It was also a pain-killing drug for earaches or toothaches, and a poultice for minor wounds, swellings, and snake bites.

Native peoples in many areas cultivated tobacco. The Karok and Yurok grew it on individually owned hilltop plots upslope from their acorn-collecting places. They cured and sold tobacco to neighbors. The Maidu of northern Sierra foothills grew it on the roofs of their semi-subterranean, earth-covered lodges.

The Changing Landscape

The displacement and subjugation of an indigenous population by an invasive, ethnically different one has occurred many times throughout the history of every habitable continent. The interaction between Native Californians and Euroamericans repeats this sad, familiar theme.

The California story of displacement, however, highlights how intimately the decline of indigenous cultures was tied to the decline of pristine vegetation. The hunter-gatherer lifeway requires ecosystems large enough to permit tribes the freedom to move with the season, harvesting a sequence of different vegetation types. As major vegetation types became more rare and endangered, so too did native cultures. Tule marsh, healthy oak woodland, riparian forest, pinyon woodland, ungrazed montane meadow, desert basin land—each a prime habitat for large human populations—have all plummeted in acreage and in quality. What was once vast and open to all has become fragmented, fenced, and privately owned. What was grazed lightly and seasonally by native animals has become overgrazed year-round by domesticated livestock. What was fished, hunted, and harvested by a few has become intensively exploited by many or entirely displaced. What flowed freely has been diverted. What was burned is now protected from fire.

Pre-European California vegetation now occupies only a small fraction of the state; so too do the Native Californians. Today land reserved for them and administered by the Bureau of Indian Affairs amounts to half a million acres—that is, one-half of one percent of the state's total land area. Approximately 50,000 people identify themselves ethnically as Native Californians, but only one-fifth of that number live on reservations.

The largest reservation is Hoopa Valley, in Humboldt County, with 86,000 acres. This area includes a typical northern California mix of Coast Range vegetation: grassland, oak woodland, riparian forest, chaparral, and mixed evergreen forest. The Hoopa Valley Reservation approximates a square, twelve miles on each side. Even in pristine times, with a non-degraded plant cover, this would not have been large enough to provide an adequate, dependable sequence of habitats. Reservation land also abuts a crowded, changed world; ecosystems within it are not autonomous from, or independent of, development which occurs at the outskirts. Some few hundred people living today may practice remnants of the hunter-gatherer lifeway and preserve elements of its culture, but an entire people can no longer live that way.

When the lifeway diminished, so did much of the ecological knowledge that Native Californians had acquired over hundreds of generations and thousands of years. Knowledge has been lost about vegetation management by fire, linkage between quality of plant cover and the health of animal populations, maintenance techniques for rare and valuable plant populations, therapeutic and nutritional values of plant species, and the human carrying capacity of the land.

Nutrients and Calories of Pine Seeds, Acorns, and Modern Wheat (adapted from Farris, 1980)

Plant Material	Protein (%)	Fat (%)	Carbohydrate (%)	Energy Content (calories/3.5 oz)
Sugar pine seed	21.4	53.6	17.5	594
Foothill pine seed	25.0	49.4	17.5	571
Coulter pine seed	25.4	51.0	14.4	574
Single-leaf pinyon pine seed	8.1	23.0	56.3	450
Valley oak acorn	4.8	18.6	65.9	440
Black oak acorn	3.8	19.8	64.8	443
Corn flour	7.8	2.6	76.8	361
Camas bulb	0.7	0.2	27.1	110
Wild onion	2.2	0.4	20.8	96
Modern wheat	13.3	2.0	71.0	352

8

RESTORING CALIFORNIA VEGETATION

Appreciation of the wilderness must be seen as recent, revolutionary, and incomplete. Ambivalence, a blend of attraction and repulsion, is still the most accurate way to characterize the present feeling toward wilderness. Wilderness has risen far on the scale of man's priorities. But the depth and intensity of previous antipathy suggests that it still has a long way to go.

Roderick Nash
Wilderness and the American Mind

AMERICAN BELIEFS, ATTITUDES, AND POLICIES TOWARD THE ENvironment are continually changing. As more of us have positive experiences in the natural environment, and as that natural environment shrinks, our society no longer fears it nor endeavors to subdue it. The closer we come to comprehending that we are an integral part of nature, the better we understand the consequences of our actions in nature. We have evolved from a society that instinctively cleared natural cover—often impoverishing the best land—to one that is beginning to conserve and enhance that cover.

Two centuries of grazing, farming, logging, mining, and building have changed California's natural cover. The net effect so far has been a reduction in biological diversity and structure and the interruption of ecological relationships that maintain balance and stability.

Until the past two decades disruption was minimally repaired at best. According to the 1970 California Environmental Quality Act (CEQA), it is the state's goal "to develop and maintain a high-quality environment," "to protect, rehabilitate, and enhance the environmental quality of the state," and "to preserve for future generations representations of all plant and animal communities." CEQA requires an evaluation of probable impacts and mitigating actions for significant environmental impacts. Habitats must be created or enhanced elsewhere to replace those degraded by development. CEQA sets the stage for going beyond the status quo; it states that both rehabilitation and enhancement must occur.

Much of the earth's surface today requires rehabilitation and enhancement, and California is an important place to begin. California's 100 million acres are only 1/360th of the world's thirty-six billion terrestrial acres, but we are entrusted

Can lush riparian forests, once common in the Central Valley, be preserved and restored?

with a landscape and a collection of organisms whose unique importance represents much more than 0.3 percent of the world's land area. Within its unique ecosystems, California has ninety-nine percent of all the redwoods (*Sequoia sempervirens*) on earth—the tallest life form; all of the giant sequoias (*Sequoiadendron gigan-teum*)—the most massive life form; and many of the bristlecone pines (*Pinus longaeva*)—the oldest tree life form. Hundreds of less dramatic plant and animal species are found in California and nowhere else. Californians need to resolve their own local environmental problems, but in so doing, they can exert leadership

in a worldwide process of healing and restoration.

There are global reasons for restoring California's plant cover. The concentration of carbon dioxide in the atmosphere has increased by about ten percent in this century alone, with a consequent rise in world temper-

ature of 1°F. Natural vegetation serves as a correcting balance to carbon dioxide released in the burning of fuel, but when natural vegetation is degraded the balance is lost and the consequences of pollution are more immediately felt. In a global context, Californians' treatment of vegetation is the same as the Brazilians' harvest of tropical forests and the clearing of woodlands and savannas by Asians and Africans.

There are sound cultural and evolutionary reasons to restore the environment. For hundreds of thousands of years the human species and its predecessors evolved in the context of wilderness; our agricultural, pastoral, and urban history

Grasslands and oak woodlands can recover from overgrazing by livestock, but not to their original condition. One year after the removal of cattle, native herbs and bunchgrasses returned to this ranch in Sonoma County, but so did many non-native herbs and annual grasses.

is, in comparison, only a moment of time. Are we genetically equipped to live successfully in an artificial, managed environment? Poet-anthropologist Loren Eisely argues that man:

> . . . must learn that, whatever his powers, he lies under the spell of a greater and a green enchantment which, try as he will, he can never avoid, however far he travels. The spell has been laid on him since the beginning of time—the spell of the natural world from which he sprang.

A green enchantment pulls at even the most urbanized populations. Some Tokyo citizens take short breaks from pressure-cooker days by sitting in womb-like chairs facing a wall-size poster of forest vegetation, while listening to sounds of nature on ear phones. The soothing impact of these imaginary nature breaks is impressive, given their artificiality. How much better must be the real thing.

Three Futures for California

There are a range of possibilities for the future of California's plant cover. The following three scenarios reflect differing human attitudes toward vegetation. The futures are not perfect visions which will become reality overnight; they will be realized gradually as our attitudes, policies, and actions slowly accumulate and become set. The choice of futures is ours.

Future I: Development with Conservation and Enhancement

In this envisioned future it is understood that regulated development, local planning, limited conservation, and mitigation of project impacts are not enough to halt cumulative, mounting environmental damage. Regulated development is combined with vigorous protection of remaining natural areas and with enhancement and healing of already degraded habitats.

Local habitat enhancement plans are developed and implemented in conformance with goals adopted at regional and statewide levels. Proposed developments are assessed for the ecological enhancements they bring as well as significant economic impacts. In practice, as well as in theory, protection is given to every aspect of the environment.

There is a formal process for the public to nominate and acquire lands for restoration. Tax funds from those who profit from land development are used for habitat restoration, the buy-out of development rights, and the compensation of previous owners. Mechanisms are created to allow private land with desirable habitat values to be traded for public land that would be developed in its place.

Long-term environmental and social benefits are goals which dominate decision making. Mitigating actions are imposed for regional and global consequences of a project, as well as for local ones. Globally, for instance, harvest of tropical wood in Brazil would be balanced by reforestation in Borneo. On a statewide basis, development in one region would be compensated by preservation in another. There is considerable public education regarding the environment, such that the importance of the California landscape would be understood

A vernal pool reserve in Sacramento County, surrounded by a recent housing development. Some reserves are so small that they cannot support the natural community or its species into the future. Better land-use planning could avoid such situations.

by everyone and ecological responsibility shared equally. All presently designated natural areas are afforded full protection.

Under this scenario market forces demand energy efficiency in all sectors. Population growth slows and demand for housing stabilizes. The economy expands with new jobs in habitat restoration, recycling, home improvement, and health care. Full recycling reduces resource consumption and waste. Networks of pedestrian paths, bikeways, and low-cost public transit encourage a shift away from automobiles and a reduction in air pollution. Industrial facilities that do not meet zero pollutant discharge standards are forced to shut down or to pay annual mitigation fees which fund restoration activities. Similar fees are collected from owners of motor vehicles (cars, planes, boats, etc.).

Improved air pollution control technology lowers levels of pollutants to the point where dangerous episodes are infrequent. Pollutants are still carried to distant areas, but the amounts are insignificant due to tight regulation. Decline in forest productivity is negligible. Collection and treatment of urban runoff dramatically reduces the degradation of downsteam bodies of water. Collection and treatment of exotic industrial wastes, recycling of hazardous waste, and construction of landfills with full linings significantly reduce groundwater pollution. Monitoring reveals a high correlation between the cleaner environment and lower incidence of birth defects, infant diarrhea, environmentally induced cancers, and the health of wildlife.

Prime agricultural lands are fully protected and preserved in perpetuity. Urbanization, judged to be incompatible with agriculture, is prohibited from adjacent areas. Conservation of irrigation water is stressed. Changes in crops and cropping schedules, and even abandonment, are considered for areas where irrigation has been historically excessive. Irrigation water subsidies are removed, so that users (and ultimately consumers of agricultural products) pay the full cost of water used, and mitigation fees go toward restoration of riparian vegetation, fisheries, and wetlands. The reuse and conservation of water lessen the need to divert stream flows and to pump ground water.

Agricultural runoff is routinely monitored, collected, and treated. Users of excess herbicides, pesticides, and fertilizers are assessed mitigation fees which pay for the processing of runoff water. Extensive pesticide monitoring programs are coupled with strict enforcement of residue laws and high penalties for infractions. The result is high-quality runoff water and low pesticide residues on crops. Universities focus agricultural research on organic farming, crop rotation, and integrated pest management. New techniques resulting from that research decrease the use of fertilizers, herbicides, and pesticides and reduce the cost of farming. Livestock are removed from riparian zones, vernal pools, and all other wetlands.

Under the Future I scenario the U.S. Forest Service develops a long-term ecological perspective. The value of old-growth forests is recognized. The multiple-use policy of the past is considered a failure. The notion of infinite renewability of forests for sustained yield is found to be fatally flawed. The past technique of assigning value to forests based solely on revenue is rejected. The dollar cost of harvested timber now includes expensive fertilization and cultivation techniques required for successful reforestation. Forests are treated as crops, not mines. Local economies are rejuvenated by revenue and jobs based in old-growth forest management, such as habitat restoration, controlled burning, selective cutting, and outdoor recreation.

On Bureau of Land Management (BLM) property, grazing is halted in many areas and permit fees are tripled in others. Inactive and unproductive mining operations are terminated. Off-road driving is banned. Critical habitats are

In some places, grazing has negative and positive effects on native vegetation. To the right of the fenceline cattle have dramatically reduced populations of native herbs. However, in the absence of grazing, non-native annual grasses can also reduce native herb populations through competition. Maintaining livestock grazing and species-rich, native grasslands will require a new kind of land management.

identified and placed in preserves. Congress finally appropriates sufficient funds to acquire staff necessary to carry out its policies. BLM corrects the cumulative damage to land resulting from decades of neglect.

Rare and endangered species are not only identified, but the vigor of their populations is periodically monitored. The conservation focus expands from the species level to the level of habitats and plant communities. Rare and endangered habitats and communities are also identified and protected. Remuneration for projects which damage threatened species or habitats is assessed on a two-for-one basis. The rate of extinction and habitat loss lessens. A process called "gap analysis," developed by Michael Scott of Hawaii and William Burley of Oregon in the 1980s, becomes a regular planning tool for conservation agencies. This technique overlays distribution maps of individual species, showing where the richest areas, most deserving of protection, are located. The process usually reveals gaps in our network of parks and preserves.

Early elements of Future I can be seen in such current policies as the protection of rare and endangered species, the establishment of wilderness areas, and the restoration of wetlands. There is already an emerging emphasis on preservation of habitats and plant communities, rather than on individual species. In some urban areas bikeways, mass transit, energy-efficient housing, and recycling are encouraged. Though limited, organic farming and integrated pest management are emerging in some agricultural areas. To fully achieve Future I, we must all recognize the importance of environmental issues and become supportive of solutions to environmental problems at all levels. We must be willing to take individual and group action to make this vision of California's future a reality.

Future II: Development with Conservation

This future represents a continuation of our present practices and attitudes. It combines regulated development with limited conservation; denigration of natural habitats continues steadily. There is strong public sentiment for reducing air and water pollution, maintaining scenic views, and protecting some wilderness areas and certain rivers. Local government regulates development so that it conforms to statewide planning and environmental protection goals. Proposed developments are evaluated for any significant environmental impacts. In theory, protection is given to every aspect of the environment, and there is a formal process for public comment and governmental review. In practice, projects are merely delayed.

Alternative sites and significant modifications to development are often proposed but seldom adopted. Short-term and long-term environmental and social costs and benefits are considered. Mitigation measures must be negotiated to reduce or compensate for significant project impacts, but our technical ability to accomplish successful mitigation is limited. Certain high quality or unique natural areas are awarded protected status, presumably forever. Overall, however, the major focus is still on development rather than on environmental quality, and the economic measures taken by government agencies reflect this bias.

Regional and global consequences of actions are addressed, but they are secondary to local concerns. There is considerable public education regarding the environment, so the ecological importance of the landscape is widely understood by decision makers.

Development is seen as an essential ingredient of local economic health. In line with this philosophy, "underutilized" land is brought to its "full economic potential," along with jobs and much needed tax revenues, as well as negative ecological consequences. Comprehensive local and county plans are drawn up, providing a certainty for growth. Growth drives the local economy, but there is a strong desire to limit heavy industry. Because of environmental restrictions on building new plants, older plants are expanded.

Efforts are made to reduce the ecological impacts of existing industrial facilities, but these are overwhelmed by population growth. Resource consumption increases, as does the creation of industrial waste products. Residential, commercial, and clean industry are preferred types of development. Nearby open lands are bought by speculators who see their potential for housing, shopping centers, and business park developments. Roads, sewers, and a fine network of electrical and communication lines are extended into these adjacent areas.

Under the Future II scenario urbanization and the desire for higher yields force agriculture to consolidate and intensify in existing districts. There is heightened competition among agricultural regions for limited water supplies. Dams are built, streams diverted, and groundwater pumping increases. Irrigation is monitored, and modest water conservation strategies are used. Crops require increased levels of fertilizers, pesticides, and herbicides. Agricultural runoff is monitored, and in some places it is sequestered in holding ponds for treatment. Rural areas are increasingly considered as sites for large power plants because of air quality problems elsewhere, the need for large volumes of cooling water, and resistance by city dwellers to urban locations.

Even with the best available technology, levels of nitrogen and sulfur oxides, ozone, hydrocarbons, carbon monoxide, and carbon dioxide continue to increase and pollution episodes are frequent. Airborne pollutants are carried to distant areas where forest productivity declines. Runoff containing oil, metals, and excess nutrients enter lakes, streams, and estuaries. There is increasing groundwater contamination by exotic compounds from "clean" industries. Wildlife reproduction declines, and environmentally induced cancers increase.

In order to rescue some timber-oriented local economies—and to supply wood for the housing industry—there is increased pressure to harvest all remaining unprotected old-growth forests, to extend forest roads, to harvest steeper slopes, to increase harvest frequency and intensity, and to replant with monocultures, using only those species which have the fastest growth rate and shortest harvest rotation time. Multiple-use policies remain in force on our national lands, and these encourage full exploitation of resources, to the detriment of the climax ecosystems. Low permit fees continue to attract and subsidize grazing, firewood cutting, and mining, in addition to timber harvest.

Demand for recreation space also increases, bringing more hikers, campers, hunters, off-roaders, and drive-through weekend tourists to natural areas. Where permitted, vacation homes are built, accompanied by environmental costs of fire control, sewage disposal, smoke from wood-burning stoves, and access roads. True wilderness becomes restricted to small islands within a vast matrix of partially degraded vegetation—museum pieces which lack the size necessary to sustain healthy populations of plants and animals.

Listed rare and endangered plant and animal species are protected, but the quality and quantity of plant cover continues to diminish. Where habitat is considered critical to threatened species, mitigation measures are taken to preserve it, including the restoration or enhancement of an equal number of acres elsewhere. Water diversion for irrigation and hydroelectric power continues to reduce natural stream flows and the biota which lives in, or next to, the water. Some flows are negotiated to satisfy sport fisheries, river rafters, and the requirements of riparian vegetation. Agricultural, logging, and grazing activities continue to encroach on riparian habitat and to change it far downstream by erosion which adds to the sediment load.

All of the elements of Future II are with us today. Future II is a continuation of present activities that allow urban areas to grow, wildlands to become degraded and exploited, and agricultural lands to shrink—while activity within them intensifies. This model historically has been the basis of the American economy, but it no longer produces a sustainable system. The environment will increasingly become a human construct, farther and farther removed from our rich, natural California cover. The passive conservation of Future II is not adequate to carry our present environment into the future.

Future III:
Development

The third future returns us to nineteenth-century practices and attitudes that marked an unbridled conversion of California's plant cover into cities, industry, forest products, and agriculture. This scenario is characterized by a retreat to poorly regulated development, state and local abdication of environmental protection roles, and low public support for conservation. Individual moneyed interests are protected to the detriment of the public good.

Under this scenario planning is restricted to the local level and is without vision. There is strong political pressure to repeal CEQA, to de-list protected species, to overturn the federal Endangered Species Act, and to limit state agency intrusion into project planning and siting. Project developers aggressively accumulate power, and the public feels unable to intervene. Decisions are based on short-term financial benefits; long-term environmental costs are rarely considered. Mitigation of environmental impacts is viewed as an unfair economic cost, and it is rarely required. Regional and global consequences of actions are seen to be outside the scope of regulatory agencies and are not considered. There is little public education regarding the environment, and the ecological importance of the deteriorating landscape is not recognized.

Growth drives the economy. Development is seen largely in terms of profits, permit fees, tax base expansion, and jobs. Easy zoning changes stimulate urbanization and industrial development. Conservation is seen as an impediment. Ecological impacts are significant only when they affect the success of development.

Nearby agricultural lands are rapidly annexed and urbanized. Farmers are forced to convert new, marginal areas to agriculture. They also intensify their activity in, and consolidate, existing agricultural land. Lower quality soils in marginal areas require increased use of fertilizers, herbicides, pesticides, and irrigation water. Runoff contains toxic levels of chemicals, but monitoring and interception are not practiced. Ground water becomes increasingly contaminated. Acceleration of groundwater pumping and diversion of surface water for agriculture permits salt water to intrude many miles from the coast. Riparian vegetation and freshwater fisheries are lost.

In Future III there is a mandate to harvest timber and to replant to monocultures with short rotation times. Multiple-use policies encourage full exploitation of federal lands. Erosion accelerates. In a desperate move to reduce the federal budget deficit, millions of acres of public Forest Service, Park Service, and BLM land are auctioned off to private ownership. Offshore drilling leases are routinely sold to oil conglomerates.

Areas inaccessible by roads become nonexistent. Plant cover is cleared, harvested, or converted to managed systems; soils are compacted; wetlands are drained and filled to be converted to other uses. Wildlife habitat is diminished in quality, and major natural cycles are disturbed. More species become rare, endangered, or extinct. As power plants, refineries, and other industries operate with minimal pollution controls, concentrations of atmospheric pollutants increase at higher rates. These pollutants are carried to distant regions, where

Conversion of agricultural land to housing near Oxnard (Ventura Co.).

Overgrazing by feral goats, deer and pigs introduced to Santa Catalina Island caused severe damage; native plant cover was widely destroyed, promoting extensive soil erosion. Removal of the exotic animals was essential for ecological restoration.

natural biotic diversity and productivity decline. Acid deposition uses up the buffering capacity of Sierran soils at an alarming rate; several lakes no longer support fish. Toxic urban runoff overwhelms lakes, streams, and estuaries far downstream. Infant diarrhea and environmentally induced cancers increase in wildlife and human populations.

Elements of Future III are with us today. These can be seen in the accelerated conversion of chaparral to residential areas or grasslands, in the promotion of off-road vehicles in deserts, in the increased harvest of old-growth timber on public lands, in the focus on private cars rather than public transport, in the redefinition of wetlands to reduce protected acreage, in resistance to the listing of new species with rare and endangered status deserving of protection, and in the encouragement of large-scale, intensive farming. In reality, Future II and III are similar; the difference between the two is the speed at which we create them.

O f the three scenarios presented, Future I may seem the least attainable, considering present economic, population, and political pressures. In fact, Future I is already happening in small but dramatic ways in California and elsewhere. The examples that follow show concretely how individuals can take the initiative in habitat restoration and preservation.

Future I Is Possible

Agriculture, urbanization, and recreation have encroached on the California desert. Irrigation projects, such as the All American and Coachella canals, provide water to farms and cities and stimulate further development. In the 1970s the U.S. Fish and Wildlife Service declared the Coachella Valley fringe-toed lizard to be a threatened species. The lizard became a focus in stopping proposed development in the area, and attitudes about development by conservationists, developers, and public agencies became polarized. The social atmosphere was tense. In 1983 The Nature Conservancy attempted a seemingly impossible task: to develop a plan for both preservation and development that would satisfy everyone. The Nature Conservancy met with all interested parties and sought their cooperation in developing and ultimately implementing a plan.

Coachella Valley Preserve

The Nature Conservancy proposed creation of a 13,000-acre Coachella Valley Preserve to protect the best remaining sand dune habitat for the fringe-toed lizard. The Coachella Valley Preserve also includes a 2,400-acre fan palm oasis. Local communities provided $2 million, the U.S. Fish and Wildlife Service $10 million, the California Wildlife Conservation Board $1 million, and the Bureau of Land Management arranged for 5 million of land exchanges. Land developers in the Coachella Valley will generate about $7 million for the sanctuary through mitigation fees paid on lands developed outside the preserve. The Coachella Valley Preserve is cooperatively managed by The Nature Conservancy and public agencies.

Creation of the Coachella Valley Preserve is a notable example of mediation and cooperation among conservationists, public agencies, and developers.

Irrigating desert lands for agriculture may produce very little economic gain for its large environmental impact. The costs of losing native species and communities, groundwater mining, surface water pollution and other impacts should be considered before wildlands are converted.

Many California cities and agricultural areas were once oak woodlands. In San Luis Obispo County, ACORN (Association for California Oak Resource News) was created to help restore oak woodlands. They have enlisted the support of the community and use volunteer labor.

In December 1985 over 200 volunteers planted more than 1,500 valley oak acorns on four acres at Lopez Lake near Arroyo Grande. This area's once abundant

Oak Woodland Restoration

Restoring valley oak woodlands and riparian forest at a Nature Conservancy preserve along the Sacramento River. Volunteers find restoration work hard but fulfilling.

oak cover had been removed over the past century. To assure success, volunteers used special planting techniques to protect oak seedlings from rodents, browsers, and drought. They planted two acorns each to five-inch diameter holes bored two feet deep. A cylinder of half-inch aviary netting lined the sides and bottom of the hole. Soil mix filled the holes and acorns were covered with two inches of soil. An additional cap of aviary netting was placed above the holes. Germination and seedling survival were very high as a result of these precautions. After six months, ninety-eight percent of the planted holes had at least one surviving seedling. The surviving saplings are now protected with wire cages four feet high and three feet wide to exclude cattle and deer.

ACORN's success shows the power of volunteer organizations to restore local plant cover. This project also shows the need to develop restoration techniques specific to the plant cover being managed.

The Nature Conservancy also has aggressively pursued restoration of riparian valley oak forest at the Cosumnes River Preserve in Sacramento County. Their approach marries privately developed technology with the labor of well organized volunteers. Over 10,000 acorns were planted, and the resulting saplings irrigated and weeded on 110 acres by supervised volunteers during 1988-89. According to

Tom Griggs, at that time manager of the Cosumnes River Preserve, the success of this project has prompted The Nature Conservancy to enlarge its restoration budget for the 1990s and to emphasize restoration as an important tool for protecting endangered vegetation. "With restoration methods that are both biologically and economically efficient, hopefully we can restore viable, naturally-functioning communities for our descendants to enjoy and to learn from in the next millennium."

The California Department of Forestry and Fire Protection recently published a detailed review of the most successful techniques that have been used in restoring valley oak in the presence of grazing. They examined several demonstration areas: Vacaville Hidden Valley in Solano County, Pepperwood Ranch Natural Preserve in Sonoma County, Wantrup Wildlife Sanctuary in Napa County, and the Cosumnes River Preserve. Appendices show how to construct protective cages and whom to contact to visit the demonstration areas (see Swiecki and Bernhardt, 1991, in the References section).

Santa Cruz Island

Prior to European colonization, Santa Cruz Island off the coast of southern California supported more than 600 plant, 200 bird, and thirteen mammal species. Early accounts show more than thirty-six species unique to the Channel Islands and eight found only on Santa Cruz Island. Several island species, such as ironwood, no longer occur on the mainland.

The seafaring Chumash Indians lived on the Channel Islands for more than 6,000 years. They gathered acorns, fruits, and bulbs; they fished and hunted. The Spanish invasion removed the Chumash to missions on the mainland. The islands remained uninhabited and undisturbed for about 100 years until herds of cattle and sheep were introduced in the mid-nineteenth century. Ample water and vegetation allowed the herds to grow to several thousand. In less than two decades they overwhelmed the islands' capacity to produce forage. First they ate the more palatable forage, and then began eating seedlings of pine, oak, and ironwood. Bite by bite they destroyed thousands of acres of natural vegetation.

In 1978 The Nature Conservancy obtained a conservation easement and ownership of ninety percent of Santa Cruz Island. The Conservancy organized groups of volunteers to repair and replace fences, confining sheep to pastures. During the next six years they conducted an intensive hunting program to clear the island of feral sheep. The island's recovery is slowly becoming apparent. The most damaged parts of the island are healing as these active conservation

Severe overgrazing and soil erosion have retarded the recovery of grassland and south coastal scrub communities on Santa Cruz Island.

measures continue. With active protection, large tracts of disturbed habitat such as Santa Cruz Island can be returned to a higher level of vegetation complexity within decades.

Arcata Marsh and Wildlife Sanctuary

Fifteen years ago state bureaucrats planned a regional sewage system for Arcata and two neighboring cities. Long recognized as sources of inadequately treated waste water pumped into Humboldt Bay, the cities were banned from further dumping unless the effluent was "enhanced." Local citizens worried that a massive project would create more problems as it solved others: a maze of sewer pipes, maintenance costs forcing a doubling of city sewage rates, fishing boat anchors snagging on submerged pipes, growth inducements along new mains. When a citizens' committee from nearby Manila sued and stalled the project for nearly two years, everybody gained time to think. Waste water has some environmental pluses, but how to satisfy legal requirements without socially and economically destroying the rural environment around the bay?

Arcata Marsh.

The proposed solution was elegantly simple: filter the post-oxidation pond water through a man-made wetland before pumping it into the bay. "Polishing" was the answer, using algae and other microbial life that cling to aquatic plant surfaces and form the base of the food chain. These small organisms remove dissolved materials such as organic pollutants in waste water.

Dubbed naive and anti-environment, Arcata's innovative approach to wastewater treatment was rejected in 1972 by the State Water Quality Control Board. Undaunted, local people drew together and fought for their project. After two years of politicking, the city won state permission for a pilot project. Finally in 1985, costing $3 million less than the original state proposal for traditional wastewater treatment, a full-sized wetland filtering waste treatment facility was in operation. "We declared victory and withdrew from the war," recalls Arcata city councilman Dan Hauser. As for war memorials, the ponds are named Hauser,

Allen, and Gearhart, members of the initial task force that created this innovative project. Today 175 acres of once abandoned industrial area support a restored wetland. Over four miles of public trails lace the marsh, and they are heavily used by residents, visitors, and school children. Its natural qualities have returned it to an important segment of the Pacific flyway for migratory wildfowl. The partially treated waste water also is used to raise juvenile Pacific salmon and trout at rates comparable to conventional fish hatcheries.

The power of the people to come together in creating innovative and cost-effective solutions is seen in Arcata. Local people, politicians, academicians, and environmentalists jointly forced the bureaucracy to be more efficient, at the same time creating an internationally known wildlife sanctuary.

If you own land you can preserve it, restore it, or manage it for maximum diversity and stability. But what if you don't own land, or don't own enough to buffer it from neighboring stresses? In answer, Eric and Steve Beckwitt sought to manage public lands: those of the USDA Forest Service. This father-and-son team shows that private citizens, well armed with facts, can have a striking influence on the management of ecosystems on vast tracts of public lands.

Saving Old-Growth Sierran Forests

Erik Beckwitt is not a trained biologist. He has a love of old-growth forests, a familiarity with them hat comes from being raised in the Sierra Nevada, and a deep well of energy that keeps his goal of preserving them focused at a constant white-hot intensity. In 1985 the Beckwitts became involved with a Sierra Club committee created to examine a draft plan published by the Forest Service for long-term management of the Tahoe National Forest. They found fault with the high intensity of logging planned for in the document, especially clear cuts proposed within old-growth stands. They spoke out via newspapers, on radio, and in meetings with school, church, and homeowner groups. As a result, the Forest Service received over 9,000 letters commenting on their forest plan, all opposing the planned logging. This level of public response to forest plans was unprecedented.

Over the next two years the Beckwitts successfully challenged several timber sales, made detailed comments on the Stanislaus National Forest draft plan, and testified before congressional committees. They also received funding from the National Audubon Society and the Save the Redwoods League to map old-growth vegetation in Sequoia and Tahoe National Forests.

The Beckwitts have found the Forest Service to be consistently responsive to private citizens when they are well informed and can document why the ecological sensitivity of certain areas makes them inappropriate for logging. Like most public agencies which manage our natural resources, the Forest Service is forced to manage incompletely known ecosystems because the funds needed to accumulate important databases have been in chronically short supply for decades. Private citizens working as volunteers for non-profit conservation

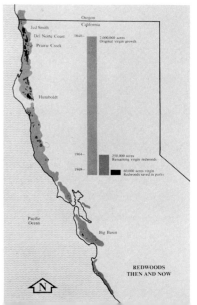

Oregon
California
Jed Smith
Del Norte Coast   1840—   7,000,000 acres
Prairie Creek   Original virgin growth

Humboldt

1964—   250,000 acres
                          Remaining virgin redwoods
1969—   60,000 acres virgin
                          Redwoods saved in parks

Pacific
Ocean

Big Basin

**REDWOODS
THEN AND NOW**

N

organizations often produce more inventory data in a short time than that which public agencies have accumulated over the course of many years. The enlarged database, as augmented by the public, then permits professional planners to re-evaluate and modify management plans.

Eric and Steve Beckwitt continue to work out of their home in the foothills east of Sacramento, mapping old-growth forests on a computer-based geographic information system (GIS). Their funding supporters now include the Brower Fund.

A dozen years ago John H. Anderson was a farmer and a primate veterinarian. Today he is also a major producer of native grass seed for potentially vast grassland restoration projects in California. His 500-acre Yolo County farm used to be completely devoted to traditional farming, but fifty acres now support tidy rows of bunchgrasses, and an additional ninety acres are being restored to native grassland cover.

John assists in regional workshops, sponsored by local agencies, which describe methods for improving wildlife habitat and restoring native vegetation on agricultural land. He is an enthusiastic and tireless promoter of bunchgrass restoration to resource managers, farmers, ranchers, and homeowners. He would like to make his new career his only one and earn a living by selling bunchgrass seed to restoration agencies and individuals. He looks forward to the time when native grass seed will be raised by farmers much as they grow other crops. To that end, he and a few other restoration pioneers created The California Native Grass Association "to develop and promote native grasses for restoration and rehabilitation of California ecosystems."

John recently became involved with a potential 40,000-acre restoration project in Tehama County, working with The Nature Conservancy on some extensively grazed private ranchland dominated by introduced annual grasses and forbs. The owner wants to restore it to pristine conditions, and John is providing advice on species to plant, planting techniques, and experimentation with cattle to determine if grazing and bunchgrasses can co-exist.

Initially John was interested simply in finding plants which would provide habitat for pheasants on his farm. Farm advisory agencies recommended two non-native perennial grasses, tall wheatgrass (*Agropyron elongatum*) and perla grass (*Phalaris tuberosa* var. *hertiglumis*). When he created hedgerows with them, he found they also suppressed weeds. He thought their weed-suppression trait would make them good roadside ground cover, so he developed a technique to propagate them which he later applied to native bunchgrasses. As a first step, an herbicide is applied soon after the first winter rains, killing germinating weeds. Then a mixture of perennial grass seed is sown and allowed to germinate during winter. In summer large weeds are spot-sprayed. By the third growing season, a dense stand of perennial grass has become established and spraying can be discontinued. Now the strip can be mowed to satisfy visibility requirements for road safety.

Perennial grasses along roadsides function as a cover or smother crop. Cover

John Anderson's Field of Dreams

Preserving some kinds of old growth forests (above left) will require new fire management policies. A combination of prescribed burns and managed natural fires should replace complete fire suppression if destructive crown fires are to be avoided and if ecological processes are to be restored.

Original, remaining, and preserved old growth redwood forests (left).

John Anderson harvests seed of native perennial bunchgrasses grown on his farm in Yolo County. The seed can be used to restore degraded, weed-dominated grasslands.

crops in agriculture have long been known to be useful in suppressing weeds. Roadside managers today, however, do not plant perennial grasses. Instead they apply herbicides as a blanket spray and later in the season scrape all dead vegetation down to ground level with a blade. "If the 800 miles of Yolo County roadsides were instead planted to perennial grass and mowed," Anderson says, "it would save $84,000 a year in herbicide and blading costs." Encouraging natural vegetation can save money.

John spends many spring weekends searching the landscape for bunchgrasses that may be adapted to local conditions. He transplants bunches or their seeds on his own land. Today he is growing more than thirty populations of fifteen different native bunchgrasses: blue grass (*Poa scabrella*), melic grass (*Melica californica*, M. *torreyana*), wild rye (*Elymus glaucus*) and creeping wild rye (*Leymus triticoides*), slender wheatgrass (*Agropyron trachycaulon* var. *majus*), deergrass (*Muhlenbergia rigens*), three-awn (*Aristida hamulosa*), June grass (*Koeleria cristata*), two species of squirreltail (*Sitanion hystrix, S. jubatum*), two species of fescue (*Festuca californica, F. idahoensis*), and four species of needle grass (*Stipa cernua, S. lemmonii, S. lepida, S. pulchra*). His conscious conservation of genetic diversity is important to successful restoration. "Local soil, geological, and topographic conditions must be matched to appropriate species and populations for best results," he says. "It might be better to leave a degraded grassland alone, than to introduce a genetically inappropriate population into it, even if that population belongs to a native California species."

John Anderson's field of dreams is bunchgrass from horizon to horizon, threading the Central Valley as it did 200 years ago.

California is rich. Economically, the annual Gross State Product places the state among the top dozen nations in the world. This is a short-term, temporary kind of richness. Ecologically, the state is also wealthy, but this wealth has been declining for two centuries. The GNP increases yearly at the expense of ecological resources.

Toward an Ecologically Rich Future

About one-fourth of our plant species are threatened with extinction, as are twenty-seven percent of our freshwater fishes, ten percent of our mammals, and six percent of our birds. Less than ten percent of such old-growth forests as riparian and coast redwood remain. Less than a fragmented ten percent of our coastal wetlands and a meager two percent of interior wetlands are still with us. The quality of most remaining cover types has been compromised by adjacent development. Oak woodlands are not reproducing, beaches are eroding because distant dams reduce sand movement into the ocean, desert scrub is damaged by off-road vehicles, grasslands are dominated by exotic species, forests have dense, flammable understories because of fire suppression policies, wetlands are no longer large and continuous enough to be valuable for waterfowl, and montane forests are weakened by ozone that has traveled hundreds of miles.

In this book we have tried to make the point that restoration of past ecological richness need not come at the expense of future economic richness. Economic prosperity requires ecological richness. Long-term benefits, as well as short-term employment opportunities, will come to human populations from restoration activities. One of the primary lessons of ecology is that everything is connected to everything else. There is no free lunch for any of us. We may be sure that there is a cost for every alteration to the environment. Alterations must be sensibly and thoughtfully weighed.

There are also ethical considerations. We have to question our present relationship with the environment. Is it right to destroy vegetation, degrading that upon which life depends? Is it right that long-term ecosystem stability be sacrificed for short-term benefits? Is it right that we exploit other species? Is it right for one generation to extinguish forever habitats that future generations will never experience?

People interested in protecting and enhancing habitat quality must scrutinize land use plans. Management proposals for public lands must be monitored for conflict with natural vegetation. Many public agencies permit logging, grazing, and mining in direct conflict with protection of the environment. These activities are not economically justified; they are heavily subsidized by taxes, and they provide limited benefits to the public. Agency plans should fairly state the economic and environmental benefits of alternative, non-destructive land uses. Often, these non-destructive uses are undervalued, poorly quantified, or ignored. It is difficult to place a price on undisturbed watershed, high biomass, photosynthesis, non-game animal habitat, rich species diversity, and ecological complexity. We have never before in human existence gone without them. We do not know the cost of their destruction, but it may be a very high one.

Although these valley oaks were not removed to develop farmland, their role in the natural community has been all but obliterated. Over time, organisms isolated from their communities can die without leaving progeny and the rich heritage of the native landscape is lost forever.

Though tremendously rewarding, conservation is hard work. It requires letters, phone calls, and visits to public officials. It requires testifying before committees and commissions. It requires organizing effective support groups and actions. It requires funds for litigation. Often the process is adversarial and results in compromise and slow change. However, if the past one hundred years of conservation efforts had not occurred in California, we would not have today's state and national parks, wilderness areas, and other protected lands.

The public has to make it known to policy makers that protection and enhancement of the environment is politically acceptable. Only major citizen involvement can make Future I a reality. At the time of this book's publication, there is evidence that some policy makers at the statewide level have begun a novel cooperative effort to conserve California's biodiversity. In September 1991 ten state and federal agencies signed a memorandum of understanding with several important features. The signers agreed to place "maintenance and enhancement of biological diversity a preeminent goal," and to join with others as members of a Biodiversity Council chaired by the secretary of the Resources Agency, currently Douglas Wheeler. The council will recommend goals, standards, and guidelines for conserving biodiversity, and it will divide the state into ten to fifteen bioregions, each with its own regional council and agenda. We have progressed, writes Marc Hoshovsky (1992), from protecting only game animals early in this century, to rare species by the 1970s, and now to biotic regions in the 1990s; from conservation initially to preservation now.

But the work is only just started. We must continue the efforts to protect remaining undisturbed areas, certainly; but we must also begin a process of restoration and enhancement of much larger areas in California, those beyond the pristine preserves. As all of California's cover regains its diversity, integrity, complexity, and stability in the coming century, so shall we.

REFERENCES AND FURTHER READING

Chapter 1

Airola, D.A. and T.C. Messick. 1987. *Sliding toward extinction: the state of California's natural heritage, 1987.* Sacramento: Jones and Stokes Associates, 105 pp.

Armstrong, W.P. 1982. Duckweeds, California's smallest wildflowers. *Fremontia* 10(3):16-22.

Axelrod, D.I. 1958. Evolution of Madro-Tertiary geoflora. *Botanical Review* 24:433-509.

Axelrod, D.I. 1976. History of the coniferous forests, California and Nevada. *University of California Publications in Botany* 70:1-62.

Bakeless, J. 1961. *The eyes of discovery.* New York: Dover, 439 pp.

Bakker, E. 1984. *An island called California,* 2nd ed. Berkeley: University of California Press, 484 pp.

Barbour, M.G. and J. Major, eds. 1988. *Terrestrial vegetation of California,* 2nd ed. Sacramento: California Native Plant Society, 1020 pp., references, and color map of pristine vegetation.

Barbour, M.G. and W.D. Billings, eds. 1988. *North American terrestrial vegetation.* New York: Cambridge University Press, 434 pp.

Beck, W.A. and Y.D. Haase. 1974. *Historical atlas of California.* University of Oklahoma Press, 101 pp., index, references.

Critchfield, W.B. 1971. *Profiles of California vegetation.* USDA Forest Service, Research Paper PSW-76, 54 pp.

Dunn, A.T. 1988. The biogeography of the California floristic province. *Fremontia* 15(4):3-9.

Durrenberger, R.W. 1974. *Patterns on the land,* 4th ed. Palo Alto, CA: National Press Books, 102 pp.

Elias, T.S., ed. 1987. *Conservation and mangement of rare and endangered plants.* Sacramento: California Native Plant Society, 630 pp.

Eyre, F.H., ed. 1980. Forest cover types of the United States and Canada. Washington, D.C.: Society of American Foresters, 148 pp., color map.

Griffin, J.R. and W.B. Critchfield. 1972. *The distribution of forest trees in California.* USDA Forest Service, Research Paper PSW-82, 114 pp.

Hill, M. 1984. *California landscape: origin and evolution.* Berkeley: University of California Press, 262 pp.

Holland, R.F. 1986. *Preliminary descriptions of the terrestrial natural communities of California.* Sacramento: California Department of Fish and Game, Nongame-Heritage Program, unpublished report, 156 pp.

Holland, R.F. 1990. *Element ranking of terrestrial natural communities.* Sacramento: California Department of Fish and Game, Nongame-Heritage Program, unpublished report, unpaged.

Hood, L., ed. 1975-82. *Inventory of California natural areas,* Volumes 1-14. California Natural Area Coordinating Council, Sonoma, CA.

Foxtail grass with dew.

Hunter, S.C. and T.E. Paysen. 1986. *Vegetation classification system for California: user's guide*. Berkeley: USDA Forest Service, General Technical Report PSW-94, 12 pp.

Johnston, V.R. In press. *The forests of California*. Berkeley: University of California Press.

Komarek, E.V., ed. 1968. *Proceedings, California Tall Timbers fire ecology conference*. Tallahassee, FL: Tall Timbers Research Station, 258 pp.

Kruckeberg, A. 1984. California serpentines. *University of California Publications in Botany* 78:1-180.

Kruckeberg, A. 1984. California's serpentine. *Fremontia* 11(4):11-17.

Latting, J., ed. 1976. *Plant communities of southern California*. Berkeley: California Native Plant Society, 164 pp.

Mason, H.L. 1969. *A flora of the marshes of California*. Berkeley: University of California Press, 878 pp.

Mason, H.L. 1970. *The scenic, scientific, and educational values of the natural landscape of California*. Sacramento, California Department of Parks and Recreation, 36 pp.

Matyas, W. and I. Parker. 1980. *CALVEG mosaic of existing California vegetation*. San Francisco: Regional Ecology Group, U.S. Forest Service, 27 pp., map.

McBride, J.R. and A. Mossadegh. 1990. Will climatic change affect our oak woodlands? *Fremontia* 18(3):55-57.

Michaelsen, J., L. Haston, and F.W. Davis. 1987. 400 years of central California precipitation variability reconstructed from tree-rings. *Water Resources Bulletin* 23:809-18.

Munz, P.A. 1974. *A flora of southern California*. Berkeley: University of California Press, 1086 pp.

Munz, P.A. and D.D. Keck. 1963. *A California flora*. Berkeley: University of California Press, 1681 pp.

Norris, R.M. and R.W. Webb. 1990. *Geology of California*, 2nd ed. New York: Wiley, 541 pp.

Ornduff, R. 1974. *Introduction to California plant life*. Berkeley: University of California Press, 152 pp.

Pemberton, R.W. 1985. Naturalized weeds and the prospects for their biological control in California. *Fremontia* 13(2):3-9.

Rejmanek, M., C.D. Thomsen, and I.D. Peters. 1991. Invasive vascular plants of California, pp. 81-101. In: R.H. Groves and F. DiCastri, eds., *Biogeography of Mediterranean invasions*. New York: Cambridge University Press.

Robinson, W.W. 1948. *Land in California*. Berkeley: University of California Press, Berkeley, 291 pp.

Sampson, A.W. and B.S. Jesperson. 1963. *California range brushlands and browse plants*. Berkeley: University of California, California Agricultural Experiment Station, Extension Service Manual No. 33, 162 pp.

Smith, J.P., Jr. and K. Berg, eds. 1988. *Inventory of rare and endangered vascular plants of California*, 4th ed. Sacramento: California Native Plant Society, 168 pp.

Smith, J.B. and D.A. Tirpak, eds. 1989. *The potential effects of global climate change on the United States. Volume I: Regional studies*. Washington, D.C., Environmental Protection Agency.

Stebbins, G.L. 1978. Why are there so many rare plants in California? *Fremontia* 5(4):6-10 and 6(1):17-20.

Sudworth, G.B. 1967. *Forest trees of the Pacific slope*. New York: Dover, 455 pp.

Van Devender, T.R. and W.G. Spaulding. 1979. Development of vegetation and climate in the southwestern United States. *Science* 204:701-10.

Chapter 2

Atwater, B.F., S.G. Conard, J.N. Dowden, C.W. Hedel, R.L. MacDonald, and W. Savage. 1979. History, landforms and vegetation of the estuary's tidal marshes, pp. 347-85. In : Conomos, T.J., ed. *San Francisco Bay: the urbanized estuary*. San Francisco: Pacific Division of American Association for the Advancement of Science.

Barbour, M.G. 1978. The effect of competition and salinity on the growth of a salt marsh plant species. *Oecologia* 37:93-99.

Barbour, M.G., Craig, R.B., Drysdale, F.R. and M.T. Ghiselin. 1973. *Coastal Ecology: Bodega Head*. Berkeley: University of California Press.

Barbour, M.G. and C.B. Davis. 1970. Salt tolerance of five California salt marsh plants. *American Midland Naturalist* 84:262-65.

Barbour, M.G. 1978. Salt spray as a microenvironmental factor in the distribution of beach plants at Point Reyes, California. *Oecologia* 32:213-24.

Barbour, M.G., T.M. DeJong, and B.M. Pavlik. 1985. Marine beach and dune plant communities, pp. 296-322. In: B.F. Chabot and H.A. Mooney, eds. *Physiological ecology of North American plant communities*. London: Chapman and Hall.

Cairns, J., ed. 1991. *Restoration of aquatic ecosystems*. Washington, D.C.: National Academy of Sciences, 485 pp.

Cooper, W.S. 1967. *Coastal dunes of California*. Boulder: Geological Society of America, Memoir 104, 131 pp.

DeJong, T.M. 1978. Comparative gas exchange and growth responses of C_3 and C_4 beach species grown at different salinities. *Oecologia* 36:59-68.

DeJong, T.M. 1979. Water and salinity relations of Californian beach species. *J. Ecol.* 67:647-63.

Faber, P.M. 1982. *Common wetland plants of coastal California: a field guide for the layman*. Mill Valley, CA: Pickleweed Press, 110 pp.

Ferreira, J.E. and K.L. Gray. 1987. *Marina State Beach dune revegetation*. Sacramento: California Department of Parks and Recreation, Natural Heritage Section, Resource Protection Division.

Harrison, A.T., E. Small, and H.A. Mooney. 1971. Drought relationships and distribution of two mediterranean-climate California plant communities. *Ecology* 52:869-75.

Johnson, D.L., ed. 1980. *The California islands*. Santa Barbara Museum of Natural History, 768 pp.

Josselyn, M., ed. 1982. *Wetland restoration and enhancement in California*. La Jolla: University of California, California Sea Grant Report T-CSGCP-007, 110 pp.

Josselyn, M. 1983. *The ecology of San Francisco Bay tidal marshes: a community profile*. Washington, D.C.: U.S. Fish and Wildlife Service, 102 pp.

Josselyn, M. and J. Buchholz. 1984. *Marsh restoration in San Francisco Bay: a guide to design and planning*. Paul F. Romberg Tiburon Center Technical Report Series Number 3, 103 pp.

Macdonald, K. 1988. Coastal salt marsh. In: M.G. Barbour and J. Major, eds. *Terrestrial vegetation of California*, 2nd ed. Sacramento: California Native Plant Society, 1020 pp.

Mahall, B.E. and R.B. Park. 1976. The ecotone between *Spartina foliosa* Trin. and *Salicornia virginica* L. in salt marshes of northern San Francisco Bay. I. Biomass and production. *J. Ecol.* 64:421-33.

Mahall, B.E. and R.B. Park. 1976. The ecotone between *Spartina foliosa* Trin. and *Salicornia virginica* L. in salt marshes of northern San Francisco Bay. II. Soil water and salinity. *J. Ecol.* 64:793-809.

Mahall, B.E. and R.B. Park. 1976. The ecotone between *Spartina foliosa* Trin. and *Salicornia virginica* L. in salt marshes of northern San Francisco Bay. III. Soil aeration and tidal immersion. *J. Ecol.* 64:811-18.

Mall, R.E. 1969. Soil-water-salt relationships of waterfowl food plants in the Suisun Marsh of California. *California Dept. of Fish and Game Wild. Bull* 1:1-59.

Malanson, G.P. 1985. Fire management in coastal sage-scrub, southern California, USA. *Environmental Conservation* 12:141-46.

Margolin, M. 1984. *The Ohlone way.* San Francisco: Heyday Books, 182 pp.

Nichols, F.H., J.E. Cloern, S.N. Luoma and D.H. Peterson. 1986. The modification of an estuary. *Science* 231:567-73.

Pavlik, B.M. 1983. Nutrient and productivity relations of the dune grasses *Ammophila arenaria* and *Elymus mollis*. I. Blade photosynthesis and nitrogen use efficiency in the laboratory and field. *Oecologia* 57:227-32.

Pavlik, B.M. 1983. Nutrient and productivity relations of the dune grasses *Ammophila arenaria* and *Elymus mollis*. II. Growth and patterns of dry matter and nitrogen allocation as influenced by nitrogen supply. *Oecologia* 57:233-38.

Pavlik, B.M. 1983. Nutrient and productivity relations of the dune grasses *Ammophila arenaria* and *Elymus mollis*. III. Spatial aspects of clonal expansion with reference to rhizome growth and the dispersal of buds. *Bulletin of the Torrey Botanical Club* 110:271-79.

Pearcy, R.W. and S.L. Ustin. 1984. Effects of salinity on growth and photosynthesis of three California tidal marsh species. *Oecologia* 62:68-73.

Seliskar, D.M. and J.L. Gallagher. 1983. *The ecology of tidal marshes of the Pacific Northwest coast: a community profile.* Washington, D.C., U.S. Fish and Wildlife Service, 65 pp.

Shuford, W.D. and I.C. Timossi. 1989. *Plant communities of Marin County.* Sacramento: California Native Plant Society, 32 pp.

Ustin, S.L., R.W. Pearcy and D.E. Bayer. 1982. Plant water relations in a San Francisco Bay salt marsh. *Bot. Gaz.* 143:368-73.

Westman, W.E. 1981. Factors influencing the distribution of species of Californian coastal sage scrub. *Ecology* 62:439-55.

Zedler, J.B. 1982. *The ecology of southern California coastal salt marshes: a community profile.* Washington, D.C.: U.S. Fish and Wildlife Service, 110 pp.

Zedler, J.B. and C.S. Nordby. 1986. *The ecology of Tijuana Estuary, California, an estuarine profile.* Washington, D.C.: U.S. Fish and Wildlife Service, Biological Report No. 85, 104 pp.

Chapter 3

Barrows, K. 1984. Old-growth Douglas-fir forests. *Fremontia* 11(4):20-23.

Carranco, L. and J.T. Labbe. 1975. *Logging the redwoods.* Caldwell, ID: Caxton Printers, 145 pp.

Dawson, W.R., J.D. Ligon, J.R. Murphy, J.P. Myers, D. Simberloff, and J. Verner. 1987. Report of the scientific advisory panel on the spotted owl. *The Condor* 89:205-29.

Faber, P.M., E. Keller, A. Sands, and B.M. Massey. 1989. *The ecology of riparian habitats of the southern Californian coastal region: a community profile.* Washington, D.C.: U.S. Fish and Wildlife Service, Biological Report No. 85, 152 pp.

Forsman, E.D., E.C. Meslow, and H.M. Wight. 1984. Distribution and biology of the spotted owl in Oregon. *Wildlife Monographs* 87:1-64.

Franklin, J.F., K. Cromack, Jr., W. Denison, A. McKee, C. Maser, J. Sedell, F. Swanson, and G. Juday. 1981. *Ecological characteristics of old-growth Douglas-fir forests.* Portland, OR: U.S. Forest Service, General Technical Report PNW-118, 48 pp.

Hyde, P. and F. Leydet. 1969. *The last redwoods.* San Francisco: Sierra Club-Ballantine Books, 160 pp.

Jenny, H., R.J. Arkley, and A.M. Schultz. 1969. The pygmy forest podsol ecosystem and its dune associates of the Mendocino coast. *Madrono* 20:60-74.

Pavlik, B.M., P.C. Muick, S. Johnson, and M. Popper. 1991. *Oaks of California.* Los Olivos, CA: Cachuma Press, 184 pp.

Sholars, R.E. 1984. The pygmy forest of Mendocino. *Fremontia* 12(3):3-8.

Strong, D.R., D. Simberloff, K.R. Dixon, T.C. Juelson, and H. Salwasser. 1987. Spotted owls. *Ecology* 68:765-79.

Waring, R.H. and J.F. Franklin. 1979. Evergreen coniferous forests of the Pacific Northwest. *Science* 204:1380-86.

Chapter 4

Allen-Diaz, B.H. and B.A. Holzman. 1991. Blue oak communities in California. *Madrono* 38:80-95.

Anderson, W. 1991. *The Sutter Buttes: a naturalist's view.*, 3rd printing. Prescott, AZ: Anderson Press, 326 pp.

Baker, H.G. 1985. What is a weed? *Fremontia* 12(4):7-12.

Barbour, M.G. and V. Whitworth. 1992. California's grassroots: native or European? *Pacific Discovery* 45(1):8-15.

Barry, W.J. 1972. *The central valley prairie.* Sacramento: California Resources Agency, Department of Parks and Recreation, 82 pp.

Bartolome, J.W. and B. Gemmill. 1981. The ecological status of *Stipa pulchra* (Poaceae) in California. *Madrono* 28:172-85.

Burcham, L.T. 1957. *California range land: an historical-ecological study of the range resource of California.* Sacramento: California Department of Natural Resources, Division of Forestry, 261 pp., map.

Burcham, L.T. 1981. California rangelands in historical perspective. *Rangelands* 3:95-104.

Conrad, C.E. and W.C. Oechel, eds. 1982. *Dynamic land management of mediterranean-type ecosystems.* Berkeley: U.S. Forest Service, General Technical Report PSW-58, 637 pp.

DiCastri, F. and H.A. Mooney, eds. 1973. *Mediterranean-type ecosystems, origin and structure.* New York: Springer.

Edwards, S.W. 1992. Observations on the prehistory and ecology of grazing in California. *Fremontia* 20(1):3-11.

Faber, P.M., ed. 1990. Year of the oak. *Fremontia* 18(3):2-112.

George, M.R., J.R. Brown, M. Robbins, and W.J. Clawson. 1990. *An evaluation of range condition assessment on California annual grassland.* Sacramento: California Department of Forestry and Fire Protection, 42 pp.

Griffin, J.R. 1971. Oak regeneration in the upper Carmel Valley, California. *Ecology* 52:862-68.

Haslam, G.W. 1990. *The other California: the great Central Valley in life and letters.* Santa Barbara, CA: Capra Press.

Haslam, G.W. 1990. Taming Rio Bravo. *Pacific Discovery*, March/April: 3-13.

Haslam, G.W. 1989. When Bakersfield was an island. *Berkeley Monthly*, March: pp. 24, 28, 29.

Hayakawa, M., ed. 1976. Vernal pools. Fremontia 4(3):3-28.

Holland, R.F. 1978. *The geographic and edaphic distribution of vernal pools in the Great Central Valley, California*. Berkeley: California Native Plant Society, 12 pp., maps.

Ikeda, D.H. and R.A. Schlising, eds. 1990. *Vernal pool plants: their habitat and biology*. Chico, CA: California State University, Studies from the Herbarium No. 8, 178 pp.

Kruckeberg, A. *et al.* 1984. The flora of California's serpentine. *Fremontia* 11(5):3-33.

Mooney, H.A. and E.L. Dunn. 1970. Convergent evolution of mediterranean- climate evergreen sclerophyll shrubs. *Evolution* 24:292-303.

Mooney, H.A. and E.L. Dunn. 1972. Land-use history of California and Chile as related to the structure of the sclerophyll scrub vegetations. *Madrono* 21:305-19.

Plumb, T.R. and N.H. Pillsbury, eds. 1987. Multiple-use management of California's hardwood resources. Berkeley: U.S. Forest Service, General Technical Report PSW-100, 462 pp.

Plumb, T.R., ed. 1980. *Ecology, management, and utilization of California oaks*. Berkeley: U.S. Forest Service, General Technical Report PSW-44, 368 pp.

Preston, W.L. 1981. *Vanishing landscapes: land and life in the Tulare Lake Basin*. Berkeley: University of California Press, 298 pp.

Rundel, P.W. 1986. Structure and function in California chaparral. *Fremontia* 14(3):3-10.

Sands, A., ed. 1977. *Riparian forests in California*. Davis: University of California, Institute of Ecology Publication No. 15, 122 pp.

Spira, T.P. and L.K. Wagner. 1983. Viability of seeds up to 211 years old extracted from adobe brick buildings of California and northern Mexico.   *American Journal of Botany* 70:303-07.

Tucker, J.M. 1983. California's native oaks. *Fremontia* 11(3):3-12.

Warner, R.E. and K. M. Hendrix, eds. 1984. *California riparian systems*. Berkeley: University of California Press, 1035 pp.

Zedler, P.H. 1987. *The ecology of southern California vernal pools: a community profile*. Washington, D.C.: U.S. Fish and Wildlife Service, 136 pp.

Chapter 5

Axelrod, D.I. 1986. The sierra redwood (*Sequoiadendron*) forest: end of a dynasty. *Geophytology* 16:25-36.

Billings, W.D. 1974. Adaptations and origins of alpine plants. *Arctic and Alpine Research* 6:129-42.

Chabot, B.F. and W.D. Billings. 1972. Origins and ecology of the Sierran alpine flora and vegetation. *Ecological Monographs* 42:163-99.

Conard, S.G. and S.R. Radosevich. 1982. Growth responses of white fir to decreased shading and root competition by montane chaparral shrubs. *Forest Science* 28:309-20.

Gains, D., M. DeDecker, and E. Bock. 1982. Owens Valley and Mono Lake. *Fremontia* 10(3):3-15.

Haller, J.R. 1962. Variation and hybridization in ponderosa and Jeffrey pines. *University of California Publications in Botany* 34:123-66.

Harvey, H.T., H.S. Shellhammer, and R.E. Stecker. 1980. *Giant sequoia ecology*. Washington, D.C.: National Park Service, Scientific Monograph Series No. 12, 182 pp.

Hill, R.B. 1986. *California mountain ranges*. Billings, MT: Falcon Press, 119 pp.

Johnston, V.R. 1970. *Sierra Nevada*. Boston: Houghton Mifflin, 281 pp.

Keeler-Wolf, T. 1990. *Ecological surveys of Forest Service Research Natural Areas in California*. Berkeley: U.S. Forest Service, General Technical Report PSW-125, 177 pp.

Kilgore, B.M. 1973. The ecological role of fire in Sierran conifer forests. *Quaternary Research* 3:496-513.

Minnich, R.A. 1987. The distribution of forest trees in northern Baja California, Mexico. *Madrono* 34:98-127.

Rundel, P.W. 1971. Community structure and stability in the giant sequoia groves of the Sierra Nevada, California. *American Midland Naturalist* 85:478-92.

VanKat, J.L. and J. Major. 1978. Vegetation change in Sequoia National Park, California. *Journal of Biogeography* 5:377-402.

Williams, W.T., M. Brady, and S.C. Willison. 1977. Air pollution damage to the forests of the Sierra Nevada Mountains of California. *Journal of the Air Pollution Control Association* 27:230-34.

Williams, W.T. 1983. Tree growth and smog disease in the forests of California: case history, ponderosa pine in the southern Sierra Nevada. *Environmental Pollution* 30:59-75.

Williams, W.T. and J.A. Williams. 1986. Effects of oxidant air pollution on needle health and annual ring width in a ponderosa pine forest. *Environmental Conservation* 13:229-34.

Chapter 6

Barbour, M.G. 1968. Germination requirements of the desert shrub *Larrea divaricata*. *Ecology* 49:915-23.

Beatley, J.C. 1966. Ecological status of introduced rome grasses (*Bromus* spp.) in desert vegetation of southern Nevada. *Ecology* 47:548-54.

Beatley, J.C. 1974. Effects of rainfall and temperature on the distribution and behavior of *Larrea tridentata* (creosote bush) in the Mojave Desert of Nevada. *Ecology* 55:245-61.

Beatley, J.C. 1974. Phenological events and their environmental triggers in Mojave Desert ecosystems. *Ecology* 55:856-63.

Benson, L. and R.A. Darrow. 1981. *Trees and shrubs of the southwestern deserts*. Tucson: University of Arizona Press, 416 pp.

Burk, J.H. 1988. Sonoran desert, pp. 869-89. In: M.G. Barbour and J. Major, eds. *Terrestrial vegetation of California*, 2nd edition. Sacramento: California Native Plant Society, 1020 pp.

Caldwell, M. 1985. Cold desert, pp. 198-212. In: B.F. Chabot and H.A. Mooney, eds. *Physiological ecology of North American vegetation*. New York: Chapman and Hall, 351 pp.

California Department of Fish and Game. 1988. *A review of the status of the desert tortoise* (*Gopherus agassizii*). Sacramento: California Resources Agency, Department of Fish and Game, Inland Fisheries Division.

Cole, K.L. and R.H. Webb. 1985. Late Holocene vegetation changes in Greenwater Valley, Mojave desert, California. *Quaternary Research* 23:227-35.

Cook, C.W. 1976. Surface-mine rehabilitation in the American west. *Environmental Conservation* 3:179-83.

Cooke, R.U. and A. Warren. 1973. *Geomorphology in deserts*. Berkeley: University of California Press, 374 pp.

DeDecker, M. 1984. *Flora of the northern Mojave desert, California.* Berkeley: California Native Plant Society, 163 pp.

Ehleringer, J. 1985. Annuals and perennials of warm deserts, pp. 162-180. In: B.F. Chabot and H.A. Mooney, eds. *Physiological ecology of North American vegetation.* New York: Chapman and Hall, 351 pp.

Hunt, C.B. 1966. *Plant ecology of Death Valley, California.* U.S. Geological Survey Professional Paper No. 509, 68 pp.

Jaeger, E.C. 1941. *Desert wildflowers.* Palo Alto, CA: Stanford University Press, 322 pp.

Johnson, H.B. 1976. Vegetation and plant communities of southern California deserts—a functional view, pp. 125-64. In: J.L. Latting, ed. *Plant communities of southern California.* Berkeley: California Native Plant Society.

Johnson, S. 1987. Can tamarisk be controlled? *Fremontia* 15:2, 19-20.

Jordan, P.W. and P.S. Nobel. 1981. Seedling establishment of *Ferocactus acanthodes* in relation to drought. *Ecology* 62:901-06.

Kay, B.L. 1979. *Summary of revegetation attempts on the second Los Angeles aqueduct.* Davis: University of California, Department of Agronomy and Range Science, Mojave Revegetation Notes No. 22, 23 pp.

Kay, B.L. 1988. Artificial and natural revegetation of the second Los Angeles aqueduct. Davis: University of California, Department of Agronomy and Range Science, Mojave Revegetation Notes No. 24, 32 pp.

Kurzius, M.A. 1981. Vegetation and flora of the Grapevine Mountains, Death Valley National Monument, California-Nevada. Las Vegas: University of Nevada, National Park Service Cooperative Research Unit, Contribution No. 017/06, 289 pp.

Mack, R.N. 1981. Invasion of *Bromus tectorum* L. into western North America: an ecological chronicle. *Agro-Ecosystems* 7:145-65.

MacMahon, J.A. 1988. Warm deserts, pp. 231-64. In: M.G. Barbour and W.D. Billings, eds. *North American terrestrial vegetation.* Cambridge University Press, 434 pp.

MacMahon, J.A. and F.H. Wagner. 1985. The Mojave, Sonoran, and Chihuahaun deserts of North America, pp. 105-202. In: M. Evenari, I. Noy-Meir, and D.W. Goodall, eds. *Hot deserts and arid shrublands.* Amsterdam: Elsevier.

Mooney, H.A., J. Ehleringer, and J.A. Berry. 1976. High photosynthetic capacity of a winter annual in Death Valley. *Science* 194:322-24.

Mulroy, T.W. and P.W. Rundel. 1977. Annual plants: Adaptations to desert environments. *Bioscience* 27:109-15.

Nobel, P.S. 1985. Desert succulents, pp. 181-97. In: B.F. Chabot and H.A. Mooney, eds. *Physiological ecology of North American vegetation.* New York: Chapman and Hall, 351 pp.

Noy-Meir, I. 1973. Desert ecosystems, environment and producers. *Annual Review of Ecology and Systematics* 4:25-51.

Pavlik, B.M. 1980. Patterns of water potential and photosynthesis of desert sand dune plants, Eureka Valley, California. *Oecologia* 46:147-54.

Pavlik, B.M. 1985. Sand dune flora of the Great Basin and Mojave deserts, California, Nevada and Oregon. *Madroño* 32:197-213.

Pavlik, B.M. 1988. Phytogeography of sand dunes in the Great Basin and Mojave deserts. *Journal of Biogeography* 16:227-38.

Rowlands, P.G. 1980. Recovery, succession and vegetation in the Mojave desert. In: P.G.

Rowlands, ed. *The effects of disturbance on desert soils, vegetation and community processes with special emphasis on off-road vehicles.* Washington, D.C.: U.S. Department of the Interior, Bureau of Land Management, Desert Plan Staff Special Publication.

Rowlands, P.G. and J.A. Adams. 1980. The effects of off-road vehicles on soils, vegetation and community processes. In: P.G. Rowlands, ed. *The effects of disturbance on desert soils, vegetation and community processes with special emphasis on off-road vehicles.* Washington, D.C.: U.S. Department of the Interior, Bureau of Land Management, Desert Plan Staff Special Publication.

Schad, J. 1988. *California deserts.* Billings, MT: Falcon Press, 120 pp.

Shreve, F. 1942. The desert vegetation of North America. *Botanical Review* 8:195-246.

Smith, S.D. and P.S. Nobel. 1986. Deserts, pp. 3-62. In: N.R. Baker and S.P. Long, eds. *Photosynthesis in Contrasting Environments.* New York: Elsevier Science.

Solbrig, O.T. and G.H. Orians. 1977. The adaptive characteristics of desert plants. *American Scientist* 65:412-21.

Spaulding, W.G. 1985. *Vegetation and climate of the last 45,000 years in the vicinity of the Nevada Test Site, south-central Nevada.* U.S. Geological Survey Professional Paper No. 1329, 55 pp.

Stebbins, R.C. 1974. Off-road vehicles and the fragile desert. *The American Biology Teacher* 36:203-34.

Steenbergh, W. F. and C.H. Lowe. 1977. *Ecology of the saguaro: II. Reproduction, germination, establishment, growth, and survival of the young plant.* National Park Service Scientific Monograph Series No. 8, 242 pp.

Thorhaug, A. 1980. Recovery patterns of restored major plant communities in the United States, pp. 113-24. In: J. Cairns, ed. *The recovery process in damaged ecosystems.* Ann Arbor Science, 167 pp.

Turner, R.M., S.M. Alcorn, G. Olin, and J.A. Booth. 1966. The influence of shade, soil and water on saguaro seedling establishment. *Botanical Gazette* 127:95-102.

Van Devender, T.R. and W.G. Spaulding. 1979. Development of vegetation and climate in the southwestern United States. *Science* 204:701-10.

Vasek, F.C. 1980. Creosote bush, long-lived clones in the Mojave desert. *American Journal of Botany* 67:246-55.

Vasek, F.C. and M.G. Barbour. 1977. Mojave desert scrub vegetation, pp. 835-67. In: M.G. Barbour and J. Major, eds. *Terrestrial vegetation of California.* New York: John Wiley & Sons, 1002 pp.

Vasek, F.C., H.B. Johnson, and D.H. Eslinger. 1975. Effects of pipeline construction on creosote bush scrub vegetation of the Mojave desert. *Madroño* 23:1-13.

Vasek, F.C., H.B. Johnson, and G. D. Brum. 1975. Effects of power transmission lines on vegetation of the Mojave desert. *Madroño* 23:114-30.

Vogl, R.J. and L.T. McHargue. 1966. Vegetation of California fan palm oases on the San Andreas fault. *Ecology* 47:532-40.

Webb, R. H., J.W. Steiger, and E. B. Newman. 1988. *The response of vegetation to disturbance in Death Valley National Monument, California.* U.S. Geological Survey Bulletin 1793, 103 pp.

Webb, R.H. and H.G. Wilshire, eds. 1983. *Environmental effects of off-road vehicle use: impacts and management in arid regions.* New York: Springer-Verlag, 534 pp.

Webb, R.H. and H.G. Wilshire. 1979. Recovery of soils and vegetation in a Mojave desert ghost town, Nevada, U.S.A. *Journal of Arid Environments* 3:291-303.

Webb, R.H. nd Stielstra. 1979. Sheep grazing effects on Mojave desert vegetation and soils. *Environmental Management* 3:517-29.

West, N. 1988. Intermountain deserts, shrub steppes and woodlands, pp. 209-30. In: M.G. Barbour and W.D. Billings, eds. *North American terrestrial vegetation*. Cambridge University Press, 434 pp.

West, N.E., K.H. Rea, and R.D. Harniss. 1979. Plant demographic studies in sagebrush-grass communities of southeastern Idaho. *Ecology* 60:376-88.

Young, J.A., R.A. Evans, and J. Major. 1972. Alien plants in the Great Basin. *Journal of Range Management* 25:194-201.

Zedler, P.H. 1981. Vegetation change in chaparral and desert communities, pp. 406-30. In: D.C. West, H.H. Shugart, and D.B. Botkin, eds. *Forest succession: concepts and applications*. New York: Springer-Verlag.

Chapter 7

Anderson, M.K. 1990. California Indian horticulture. *Fremontia* 18(2):7-14.

Anderson, M.K. 1991. California Indian horticulture: management and use of redbud by the southern Sierra Miwok. *Journal of Ethnobiology* 11:145-57.

Anderson, M.K. and G.P. Nabhan. 1991. Gardeners in Eden. *Wilderness* (Fall):27-30.

Balls, E.K. 1972. *Early uses of California plants*. Berkeley: University of California Press, 103 pp.

Barrows, D.P. 1967. *Ethnobotany of the Coahuilla Indians of southern California*. Banning, CA: Malki Museum Press, 82 pp.

Bettinger, R.L. 1976. The development of pinyon exploitation in central eastern California. *Journal of California Anthropology* 3:81-95.

Chestnut, V.K. 1902. *Plants used by the Indians of Mendocino County, California*. Contributions from the U.S. National Herbarium, Volume VII. Reprinted 1990 by the Mendocino County Historical Society, Fort Bragg, California, 127 pp.

Clarke, C.B. 1977. *Edible and useful plants of California*. Berkeley: University of California Press, 280 pp.

Cook, S.F. 1973. The aboriginal population of Upper California. In: R.F. Heizer and M.A. Whipple, eds. *The California Indians*. Berkeley: University of California Press.

Cook, S.F. 1976. *The conflict between the California Indian and white civilization*. Berkeley: University of California Press, 522 pp.

Cook, S.F. 1978. Historical demography, pp. 91-8. In: R.F. Heizer, ed. *Handbook of North American Indians*, volume 8. Washington, D.C.: Smithsonian Institution.

d'Azevido, W.L., ed. 1986. *Handbook of North American Indians*, volume 11. Washington, D.C.: Smithsonsian Institution.

Elsasser, A.B. 1981. Notes on Yana ethnobotany. *Journal of California and Great Basin Anthropology* 3(1):69-77.

Farris, G. 1980. A reassessment of the nutritional value of *Pinus monophylla*. *Journal of California and Great Basin Anthropology* 2(1):132- 36.

Heizer, R.F., ed. 1978. *Handbook of North American Indians*, volume 8. Washington, D.C.: Smithsonian Institution, 800 pp.

Heizer, R.F. and T. Kroeber, eds. 1979. *Ishi: the last Yahi*. Berkeley: University of California Press, 242 pp.

Heizer, R.F. and M.A. Whipple, eds. 1971. *The California Indians*, 2nd ed. Berkeley: University of California Press, 619 pp.

Hinton, L. 1975. Notes on La Huerta Diegueno ethnobotany. *Journal of California and Great Basin Anthropology* 2(2):214-22.

Kirk, D.R. 1975. *Wild edible plants of western North America.* Happy Camp, CA: Naturegraph.

Kroeber, T. and R.F. Heizer. 1968. *Almost ancestors: the first Californians.* New York: Sierra Club/Ballantine Books, 168 pp.

Lawton, H.W., P.J. Wilke, M. DeDecker, and W.M. Mason. 1976. Agriculture among the Paiute of Owens Valley. *Journal of California Anthropology* 3:13-50.

Lee, R.B. and I. Devore. 1979. *Man the hunter.* New York: Aldine, 415 pp.

Margolin, M., ed. 1981. *The way we lived.* San Francisco: Heyday Books, 209 pp.

Margolin, M., ed., 1987—. *News from native California.* Newsletter published every two months by Heyday Books, San Francisco, CA.

Margolin, M. 1989. *Monterey in 1786: the journals of Jean Francois de La Perouse.* Berkeley: Heyday Books, 111 pp.

Merriam, C.H. 1905. The Indian population of California. *American Anthropologist* 7:594-606.

Merrill, R.E. 1923. Plants used in basketry by the California Indians. *University of California Publications in American Archaeology and Ethnology* 20:215-42.

Murphey, E.V.A. 1959. *Indian uses of native plants.* Reprinted 1987 by the Mendocino County Historical Society, Fort Bragg.

Purdy, C. 1902. *Pomo Indian baskets and their makers.* Reprinted 1960 by the Mendocino County Historical Society, Ukiah, California, 44 pp.

Schenk, S.M. and E.W. Gifford. 1952. Karok ethnobotany. *University of California Archaeological Records* 13:377-92.

Scully, V. 1972. *A treasury of American Indian herbs.* New York: Crown, 306 pp.

Shipley, W.F. 1978. Native languages of California, pp. 80-90. In: R.F. Heizer, ed. *Handbook of North American Indians*, volume 8. Washington, D.C.: Smithsonian Institution.

Thompson, S. and M. Thompson. 1972. *Wild food plants of the Sierra.* San Francisco: Dragtooth Press, 186 pp.

Ubelaker, D.H. 1992. North American census, 1492. *Pacific Discovery* 45(1):32-35.

Chapter 8

Anonymous. 1979. *California's forest resources.* Sacramento: California Department of Forestry, 352 pp., appendix.

Anonymous. 1981. *CEQA: The California Environmental Quality Act.* Sacramento: California Office of Planning and Research, 141 pp.

Anonymous. Coachella Valley Preserve. *The Nature Conservancy News*, Dec86/Jan87, p. 11.

Berg, K. and D. Ikeda. In press. *California's vanishing flora.* Sacramento: California Department of Fish and Game, California Academy of Sciences, and California Native Plant Society.

Berger, J.J. 1985. *Restoring the earth: how Americans are working to renew our damaged ecosystems.* New York: Knopf, 241 pp.

Brinck, P., L.M. Nilsson, and U. Svedin. 1988. Ecosystem redevelopment. *Ambio* 17(2):84-89.

Buschbacher, R. J. 1986. Tropical deforestation and pasture development. *BioScience* 36(1):22-28.

Carter, H.O., R. Coppock, L. Kennedy, C. Nuckton, and J. Spezia. 1991. Keeping the valley green: a public policy challenge. *California Agriculture* 45(3):10-14.

Dasmann, R.F. 1968. *Environmental conservation*. New York: Wiley, 375 pp.

Devall, B. and G. Sessions. 1985. *Deep ecology*. Salt Lake City: Peregrine Smith Books, 267 pp.

Dunning, H.C. 1982. California water rights law: the need for change. *Fremontia* 10(1):14-17.

Ewing, R.A., ed. 1988. *California's forests and rangelands: growing conflict over changing land uses*. Sacramento: California Department of Forestry and Fire Protection, 348 pp., appendices.

Glasby, G.P. 1986. Modification of the environment in New Zealand. *Ambio* 15(5):266-71.

Grossi, R., W. Shafroth, J. Hart, and M.J. Singer. 1987. California's shrinking farmland. *California Agriculture* 41(7/8):22-24.

Hansen, B. 1987. Santa Cruz: an island reborn. *The Nature Conservancy News*, June/July, pp. 9-14.

Hoshovsky, M. 1992. Developing partnerships in conserving California's biological diversity. *Fremontia* 20(1):19-23.

Jackson, W. 1980. *New roots for agriculture*. New York: Friends of the Earth, 155 pp.

Jensen, D.B., M. Torn, and J. Harte. 1990. *In our own hands: a strategy for conserving biological diversity in California*. Berkeley: University of California, California Policy Seminar, 184 pp., appendices.

Jones and Stokes Associates. 1987. *Sliding toward extinction: the state of California's natural heritage*. Commissioned by The California Nature Conservancy, Sacramento, 105 pp., appendices.

Jordan, W.R. III., M.E. Gilpin, and J.D. Aber, eds. 1988. *Restoration ecology*. New York: Cambridge University Press, 352 pp.

Josselyn, M. (ed.). 1982. *Wetland restoration and enhancement in California*. La Jolla: University of California, Sea Grant College Program, 110 pp.

Knox, J.B. and A.P. Scheuring, eds. 1991. *Global climate change and California*. Berkeley: University of California Press, 184 pp.

Lacey, L., ed. 1990—. *Growing Native*. Newsletter issued six times a year by Growing Native Research Institute, Berkeley, CA.

Lovelock, J.E. 1979. *Gaia: a new look at life on earth*. New York: Oxford University Press, 157 pp.

Nash, R. 1973. *Wilderness and the American mind*. Yale University Press, 425 pp.

Niesen, T. and M. Josselyn, eds. 1981. *The Hayward regional shoreline marsh restoration*. San Francisco State University, Department of Biological Sciences, Technical Report No. 1.

Petersen, R.C., Jr., B.L. Madsen, M.A. Wilzbach, C.H.D. Magadza, A. Paarlberg, A. Kullberg, and K.W. Cummins. 1987. Stream management: emerging global similarities. *Ambio* 16(4):166-79.

Snyder, G. 1990. *The practice of the wild*. San Francisco, CA: North Point Press, 190 pp.

Steinhart, P. 1990. *California's wild heritage: threatened and endangered animals in the golden state*. Sacramento: California Department of Fish and Game, California Academy of Sciences, and Sierra Club Books, 108 pp.

Swiecki, T.J. and E.A. Barnhardt. 1991. *Minimum input techniques for restoring valley oaks on hardwood rangeland*. Sacramento: California Department of Forestry and Fire Protection, 79 pp.

Thirgood, J.V. 1981. *Man and the Mediterranean forest: a history of resource depletion*. New York: Academic Press, 194 pp.

Trumbly, J.M. and K.L. Gray. 1984. The California wilderness preservation system. *Natural Areas Journal* 4(4):29-35.

White, J.E., ed. 1991. California: the endangered dream. *Time Magazine* 138(20):32-110.

GLOSSARY

Our intent has been to tell the story of California's plant cover with minimum use of the specialized vocabulary of professional ecologists, biologists, and land use managers. Nevertheless, we did include about 100 technical terms. Most terms are explained in the text where they are first introduced, but here they are again, arranged alphabetically and briefly defined.

acid rain, acid deposition The falling to earth of acidic liquids (rain, snow, fog) or acidic particles created by pollutants released to the atmosphere, usually some considerable distance away.

allelopathy The inhibition of one plant by another, caused by the addition of some metabolic byproduct introduced into the environment. This chemical substance, released by the leaves or roots of one species, reduces germination, growth, or vigor of other nearby species.

alpine A climatic zone (or vegetation growing in that zone) at high elevation where length and warmth of the growing season are insufficient to support tree growth. A similar climate and vegetation (called **arctic** or **polar tundra**) occur at low elevations at polar latitudes.

annual plant A plant whose life cycle is shorter than one year. See also **ephemeral plant.**

arroyo A desert waterway, dry most of the year but filled with running water after rain storms. Supports riparian vegetation. Also called a **wash.**

bajada A gentle desert slope, created by the erosion of coarse material from mountain slopes above.

basin A desert depression without external drainage. Its soil is fine-textured and saline, and its low elevation permits cold air to collect. Also called a **playa, sink,** or **dry lake bed.**

biological control The suppression of a pest organism by the introduction of another organism which selectively attacks only the target pest.

biomass The weight of organisms which live in a particular habitat, as in plant biomass or above-ground biomass. **Biomass conversion** is the harvest of plant biomass for burning, to provide heat or electrical energy.

bulb A specialized, short underground stem, produced by some non-woody species, such as wild onion. Bulbs store carbohydrates and are capable of vegetative reproduction.

bunchgrass A perennial grass which has no underground system for vegetative spread. All the buds are located on the base of the stems, at ground level.

Californian Relating to the Californian floristic province; plants or environment that are characteristic of that province, but that may extend beyond the borders of the state of California, as, for example, into Baja California.

carbohydrate A molecule (a chemical) that is made up of carbon, hydrogen, and oxygen.

Examples include sugar, starch, and the waxy cuticle on the surface of leaves.

chaparral Shrub-dominated vegetation which grows at low elevations away from the immediate coast. Manzanita, ceanothus, oak, and chamise are characteristic plants. Canopy cover by evergreen shrubs tends to be 100 percent. A type of **scrub** vegetation. **Montane chaparral** occurs at higher elevations and consists of different species.

clear cut A forest harvest technique in which all trees are harvested (or destroyed if too small to be of commercial value) at one time.

closed-cone conifer A conifer (pines and cypresses in California) which retains its cones attached to branches, even after the seeds inside have matured. The attached cones do not open immediately when mature, but instead open only with age, attack by animals, or exposure to heat.

cold air drainage The movement of cold, dense air downslope into a canyon or basin. Usually occurs at night when mixing winds are quiet.

community All of the species which share a given habitat. Each community has its own characteristic species composition, architecture, and habitat, and each is named after the dominant species, as a ponderosa pine forest or a chamise chaparral or a needlegrass prairie. Several communities may belong to the same regional vegetation type.

competition An interaction between neighboring organisms which both require the same resource (light, water, nitrogen, etc.). If that resource is in limited supply, the competing organisms grow or reproduce less successfully than if only one were present.

convergent evolution The process of different species becoming similar over a long period of evolutionary time.

coppice To regularly prune a plant in such a way as to promote the regrowth of many slender shoots at a reachable height.

corm A short underground stem, similar to a bulb in that it stores carbohydrates and is capable of vegetative reproduction.

cover The portion of ground covered (shaded) by any one plant species or by all plant species in a given habitat. Chaparral plants, for example, collectively cover 100 percent of the ground, and chamise plants within that vegetation might cover 40 percent of the ground.

cushion plant An alpine plant which has a well developed root system but a short, compact shoot system with small leaves.

cytoplasm The living interior part of a cell, with qualities of both a viscous fluid and a gel.

density The number of plants of one species or of all species in a given habitat; for example, 100 blue oak trees per acre in a foothill woodland.

deciduous plant A perennial plant which loses all of its leaves at the same time. **Winter-deciduous** species lose their leaves in fall, whereas **drought-deciduous** species lose their leaves following the end of a rainy period.

desert A region with an arid climate, typically receiving less than ten inches of annual rainfall. Desert vegetation (**desert scrub**) is dominated by shrubs which altogether provide less than 50 percent cover.

cold deserts are high enough in elevation to experience snow and frost in winter; **hot deserts** are low-elevation deserts with virtually no frost; **warm deserts** are intermediate.

dominant A plant species or plant growth form that characterizes a particular habitat because it is more abundant or contributes more cover than any other species or growth form. A smaller species which lives in the shade of the dominant species is called a **subdominant** or **understory** species. Any species which shares the habitat with the dominant species, regardless of its size, is called an **associated** species.

ecosystem For our purposes here, a regional unit consisting of all the plants, animals, and microbial species together with their non-living environment (soil, climate, etc.).

ecotone A meeting ground between two habitats or two environments; a transitional place.

endemic Restricted in distribution. See **native**.

environment Everything that affects an organism in its habitat: other organisms, sunlight, soil, rainfall, temperature, etc.

ephemeral An annual plant; a plant whose life span is much shorter than one year. Desert ephemerals, for example, live from six weeks to eight months, depending on growing conditions. **Winter ephemerals** germinate in winter; **summer ephemerals** germinate in summer. Also called **annual** plants.

evergreen plant A perennial plant which does not shed all its leaves at the same time; consequently, some leaves are present throughout the year. An individual leaf, of course, does not last through the entire life span of the plant; typical leaf life spans are from one to thirty years.

fire Natural fires occur in **fire climates** and are started by dry lightning strikes, not by humans. Natural fires have predictable **frequencies** or **return periods**, the number of years between fires which burn the same acre of vegetation. **Ground fires**, or surface fires, burn only low vegetation and litter, whereas **crown fires** are hotter and consume all above-ground vegetation in an area. Trees which survive a fire are often scarred when a strip of the trunk is killed, and the scar is later buried by new growth in such a way that the date of the **fire scar** can be determined by counting tree growth rings back from the present trunk surface. Species which require fire to complete their life cycle or to maintain their vigor are **fire-dependent**. The control of natural fire by humans is **fire suppression**.

flora All of the plant species within an area. The flora of California consists of 5,000 native species of higher plants.

foredune A sand dune which fronts the ocean and lies just behind the beach. It is a boundary between the beach and the interior dunes, and is created by the sand-stilling qualities of perennial dune grasses or broadleaved herbs.

forest Vegetation characterized by trees growing close enough that their canopies often touch and collectively cover more than 60 percent of the ground. **Woodland** canopies, in contrast, cover 30 to 60 percent of the ground. **Savanna** canopies cover less than 30 percent of the ground. **Old-growth forests** are mature and have been undisturbed for centuries, whereas **second-growth forests** are young and still in the process of recovering from disturbance.

fruit A technical botanical term. The seed-containing structure of flowering plants. It can be fleshy, hard, thick, thin, or papery, and need not be edible. Its function is to protect the seed and to aid in its dispersal away from the parent plant.

genus A classification category just above the species. The scientific name of an organism includes the genus followed by the **species**. *Pinus ponderosa* is the scientific name of ponderosa pine; *Pinus* is the genus name and all other pine species share that same genus name. Plural of genus is **genera**, whereas the term species is both singular and plural.

geoflora A flora from the geologic past, as revealed by the fossil record. Geofloras important in California's past include the **Neotropical-Tertiary** (warm and humid tropical), the **Arcto-Tertiary** (cold and temperate), and the **Madro-Tertiary** (warm and semi-arid).

geologic time Refers to the scale of time over which the solid earth has existed, that is, the past four to five billion years. Geologic time is broken down into epochs, eras, and periods. The remains of past surfaces, as buried strata of rock, form the **geologic record**. The remains of past plants and animals, as fossils, comprise the **fossil record**.

global warming A warming of the entire earth that is hypothesized by some scientists to be occurring now and thought likely to continue into the next century. One of the causes of global warming is thought to be the increasing concentration of carbon dioxide in the atmosphere which has been released as a pollutant from the burning of fuel.

grassland A vegetation type dominated by annual or perennial grasses. Also called a **prairie**.

growth form The shape of a plant, as a coniferous tree, a sclerophyllous shrub, or a broadleaved herb; and the behavior of that plant, as annual, perennial, parasitic, evergreen, drought-deciduous, etc. Also called **life form**.

hardpan A compacted layer of soil, sometimes several feet thick, located below the surface. Hardpan is difficult for roots and percolating soil water to penetrate. Hardpan due mainly to clay accumulation is called a **claypan.herbaceous plant** A non-woody plant, annual or perennial. If perennial, the above-ground parts die back each year, but below-ground parts remain alive.

herbicide A chemical which inhibits or deranges the metabolism and growth of plants. Generally, herbicides are created or applied in such a way that they are selective, affecting target species and leaving others unaffected.

herbivore An animal, such as a bird, insect, or mammal, which feeds on plants. Plants may possess distasteful or toxic chemicals or hard tissues which deter herbivores from feeding; these traits are part of the plant's **herbivore defense**.

hybrid The offspring of a cross between two different species. Oracle oak is a hybrid of California black oak and interior live oak parents.

intertidal The zone along the shore which is alternately inundated and exposed by high and low tides. The zone below low tide, which is never exposed, is the **subtidal**.

marsh A wetland which experiences standing water at least part of the year. The salinity of that water determines whether the marsh is a **salt marsh**, a **brackish marsh**, or a **freshwater marsh**. Salt marshes may be *coastal (tidal)* or **interior**.

mediterranean climate A climate which occurs around the rim of the Mediterranean Sea and four other places in the world, including much of California. Characterized by hot, dry summers and cool, wet winters.

microclimate, microenvironment The climate or environment close to the ground or near an organism. Differs from the regional climate or environment.

montane Refers to zones, vegetation, or climate of middle elevations in mountains. The **lower montane zone** in California is characterized by such communities as mixed conifer forest, mixed evergreen forest, and east-side Jeffrey pine forest. The **upper montane zone** is at a higher elevation and includes red fir and lodgepole pine communities. The **foothill zone** lies below the montane; the **subalpine zone**, where trees reach their elevational limit, lies above the montane.

Native Californians Peoples present in California prior to European contact. A subset of **Native American.**

native species For our purposes here, any organism present in California prior to European contact. Native plant species can be widespread, occurring in California and elsewhere, or they can be **endemic**, found only in California.

nitrogen fixation The conversion of atmospheric nitrogen gas into ammonium and its attachment to organic molecules within a cell. Only certain bacteria are able to accomplish this, and some grow in association with such plants as legumes, alders, and fungi.

ozone A reactive molecule consisting of three oxygen atoms (O_3). Ozone is formed naturally by cosmic rays in the upper atmosphere, and there it plays a protective role for life below by absorbing harmful ultraviolet radiation from the sun. Nearer the earth's surface, ozone is a byproduct of car exhaust and is harmful to plants and animals. In ponderosa pine, for example, it causes yellow lesions (mottles) on needles, early needle drop, depressed photosynthesis, and reduced trunk growth. This disease is called **ozone mottle** of pine.

Pacific flyway A major north-south migratory route for many bird species. One branch of this broad pathway passes over California, and it is utilized by birds which require wetlands as habitat.

parasite An organism which gains nourishment from a living host, generally by consuming it from within. Disease-causing microbes are parasites, as are mistletoe and dodder plant species, and animals such as ticks and mosquitoes.

perennial plant A plant which lives for more than one or two years. Perennials may be **woody** (trees, shrubs, vines) or non-woody **perennial herbs.**

pH The concentration of hydrogen in solution. **Acid** solutions have pH values below 7, whereas **basic** solutions have pH values greater than 7. Pure rainwater has a pH between 5 and 6 because natural gases in the atmosphere dissolve in rainwater and create weak acids. Industrial pollutants can create more intensive acidities below pH 4. See also **acid rain.**

photosynthesis The combination of carbon dioxide with water, in the presence of sunlight and certain enzymes, to ultimately produce the carbohydrate sugar. Several different metabolic pathways of sugar production are known: C_3, C_4, and **CAM**. Most plants have C_3 photosynthesis. Many plants of saline, sunny, dry, and/or hot habitats have C_4 photosynthesis. Certain succulent plants in arid or semi-arid habitats have CAM photosynthesis.

phreatophyte A plant whose root system is in constant contact with ground water. Typically found in riparian habitats.

physiological drought The inability of plants to take up water from a saline soil, even though the soil is wet.

pollutant A chemical or physical byproduct of human manufacture which interferes with the natural functions of organisms or ecosystems.

population A local cluster of individual organisms all belonging to the same species. A **species**, then, is typically made up of many populations scattered across the landscape. Rare species contain few populations, whereas widespread species contain many populations.

prescribed burning The purposeful setting afire of vegetation in an attempt to imitate the natural fire regime.

pristine For our purposes here, the nature of California's ecosystems prior to European contact. See also **native**.

rainshadow A relatively dry region located in the lee of a mountain range. Air will lose most of its moisture as it rises up the exposed side of a range and will be unable to contribute much moisture as it descends down the lee side (the rainshadow region).

reproduction The production of offspring. Higher plants produce seeds by **sexual reproduction**. Reproduction by rhizomes, stump sprouts, bulbs, tubers, corms, runners, and stolons is **asexual (vegetative) reproduction** and it produces offspring genetically identical to the parent. Genetically identical individuals are **clones**.

respiration The metabolic digestion of carbohydrates, such as sugar, which releases carbon dioxide, water, heat, and chemical energy as end products. The opposite of photosynthesis. Plants grow only when the rate of photosynthesis exceeds the rate of respiration. **rhizome** An elongated, horizontal underground stem capable of vegetative reproduction.

riparian The habitat, zone, or vegetation which occurs along the banks of arroyos, basins, streams, rivers, or lakes. Riparian trees and shrubs are typically **phreatophytes**.

salt A technical term for a group of chemicals (molecules) which dissolve in water to yield positively and negatively charged portions. Table salt (sodium chloride) is one example. **Salt-tolerant plants (halophytes)** tolerate higher concentrations of salt than do normal, **salt-sensitive plants (glycophytes)**.

savanna A grassland with trees regularly present, but scattered so that their canopies cover less than 30 percent of the ground. See also **forest**.

sclerophyll Literally, a hard leaf. A leaf on an evergreen plant, usually small in area, often spiny, with a thick cuticle and other features that make it drought tolerant.

scrub Vegetation dominated by shrubs. Examples include desert scrub, chaparral, dune scrub, and coastal scrub, all of which differ in species composition, habitat, and community architecture.

select cut A forest harvest technique in which a fraction of the trees are harvested at any one time, but—over the course of repeated cutting at intervals—the entire forest is eventually cut.

serpentine Soil derived from **serpentinite** rock, which is low in calcium, nitrogen, and phosporus and high in magnesium, chromium, and nickel. **soil** The terrestrial surface of the earth which has been modified by weather, plants, animals, and microbes. Soil is typically only a few feet deep and develops slowly over the course of thousands of years. Soils can be classified on the basis of their unique sequence of layers of different color, texture, and chemistry.

species A classification category below the genus rank. See also **genus** and **population**. Abbreviated **sp** (singular) or **spp** (plural).

stolon An elongated, horizontal, above-ground stem capable of vegetative reproduction. Also called a **runner**.

stomate A microscopic pore in the surface of leaves and other above-ground plant parts. Special cells which line the pore can open and close it. Plants with C_3 or C_4 photosynthesis open their stomates by day and close them by night; plants with CAM photosynthesis close their stomates by day and open them by night. Carbon dioxide enters and water vapor exits the plant through stomates.

stump-sprouting The production of new shoots from the base of a shrub stem or the base of a tree trunk. Sprouting does not ordinarily occur until the top has been killed by cutting or fire. Only certain California shrubs and trees remain alive below ground when their tops are killed, and only these species are capable of stump-sprouting. Also called **crown-sprouting**.

succulent A plant or tissue which stores water inside enlarged cells. Some succulent plants accumulate salts, and some conduct CAM photosynthesis. All succulent plants are relatively drought tolerant.

swale A wind-eroded depression in a coastal dune area, bringing the sand surface down close enough to a water table to create a localized wetland.

transpiration The loss of water vapor from plants through stomates.

tuber The swollen tip of a rhizome which stores carbohydrates and is capable of vegetative reproduction. A potato is a tuber.

tule A collective term for a group of tall, leafless freshwater marsh species in the genus *Scirpus*. Also a name for the marsh itself.

type conversion A management procedure which replaces one vegetation type with another, as the replacement of chaparral with grassland.

vacuole A membrane-bounded sac within a cell. Salts and other materials, dissolved in water, fill the vacuole and are in this way separated from the cell's cytoplasm. Succulent plants have unusually large vacuoles.

variety A classification category below the species rank. Often applied to those members of a species which occupy a regional portion of the species' range and which exhibit some minor differences from the rest of the species. May be equivalent to subspecies. Abbreviated **var** or **ssp**.

vegetation The plant cover of a region, characterized by the growth forms of its dominant plants. Vegetation type names include both geographic and growth form terms, as coastal salt marsh, montane conifer forest, and cold desert scrub. Any one vegetation type contains many communities. Vegetation types or communities which are in the process of recovering from disturbance are said to be **successional**, whereas those which are stable are said to be **climax**.

vegetative reproduction Asexual reproduction. See also **reproduction**.

vernal pool A shallow depression, generally restricted to low elevations, which collects fresh water in winter, dries in spring, and supports populations of endemic vernal pool species.

wetland By law, describes a habitat which has soil saturated within eighteen inches of the surface for at least one week of the year, and which supports certain **flood-tolerant** plant species which are wetland indicators.

woodland A vegetation type dominated by trees whose canopies collectively cover 30 to 60 percent of the ground. See also **forest** and **savanna**.

Xerothermic A period of time, 5,000 to 8,000 years ago, when temperatures in California were several degrees warmer than they are today.

INDEX OF PLANT NAMES

All plants mentioned in the text are listed here alphabetically by common name. The choice of common name is often problemmatic, because there is no standard list or even standard spelling (are the words separate, hyphenated, combined?). In those cases where more than one common name exists for a plant, we have used our best judgement in selecting the one most widely adopted. We do cross-list some of the other common names. The common name is alphabetized below by the first word. For example, California fan palm is found under "C," not "f" or "p"; tree tobacco is found under "tree," not "tobacco."

The scientific name, in parentheses, follows the common name. The first time a plant is mentioned in any chapter, its common name is followed by this scientific name. We followed Munz and Keck for scientific names in the text, as it was the most current state-wide reference available at the time we wrote the book. However, a new reference was published in early 1993—the *Jepson Manual*—and we have decided to include its new name changes in our index. New Jepson names appear in **boldface**, after the Munz and Keck name. We thank Barbara Leitner for checking our species list against galleys of the *Jepson Manual*, and the following individuals who compiled the final version of both this index and the general index: Rachel Aptekar, Jim Bouldin, Rob Fernau, Julie Oliver, Robert Rhode, Ed Royce, and Michelle Stevens.

A

agave (*Agave utahensis*, *A. deserti*, *A. shawii*) 38, 152, 177-178

alder (*Alnus* spp) 73, 75

algae (collective name for many taxa in several divisions of lower plants) 41, 167, 198

alkali bulrush (*Scirpus robustus*) 45

alkali dropseed, alkali sacaton (*Sporobolus airoides*) 147, 154

alkali heath (*Frankenia grandifolia* = **F. salina**) 43

alkali saltgrass (*Distichlis spicata* var. *stricta* = **D. spicata**) 147, 154

allscale, see desert saltbush

alpine buttercup (*Ranunculus eschscholtzii*) 119

alpine daisy (*Aster alpigenus* ssp. *andersonii*) 119

alpine everlasting (*Antennaria alpina*) 119

alpine gold (*Hulsea algida*) 119

angelica (*Angelica* spp) 179

annual pickleweed (*Salicornia bigelovii*) 42

antelope bush, see bitterbrush

apricot mallow, see globemallow

Arizona lupine (*Lupinus arizonicus*) 154

arrow-grass (*Triglochin concinnum* = *T.*

concinna, *T. maritima*) 43

arroyo willow (*Salix lasiolepis*) 31, 74

ash (*Fraxinus* spp) 73, 97

aspen, see quaking aspen

B

baby blue-eyes (*Nemophila menziesii*) 34

Baltic rush (*Juncus balticus*) 45

barley (*Hordeum* spp) 36

basin sagebrush (*Artemisia tridentata*) 134, 135, 137, 141, 179

basin wildrye (*Elymus cinereus* = **Leymus cinereus**) 136

bay, see California bay

beach-bur, see silver beachweed

beach evening primrose

Manzanita bark.

GENERAL INDEX

ABOUT THE AUTHORS

Michael Barbour has been a member of the Botany Department at the University of California, Davis for 25 years. He is nationally recognized for the quality of his teaching—in plant biology and ecology—and for his research publications of the vegetation of California and Norht America. Michael and his wife, Valerie Whitworth, live among the orchards of Winters, CA, at the western edge of the Sacramento Valley. They share a strong interest in the conservation of California's landscapes and they plan to write on that subject for school-age children. Dr. Barbour has been working with the California Native Plant Scoiety on a project to standardize the classification of California's plant communities as one step in a process of ultimately obtaining legal protection for endangered vegetation.

Bruce Pavlik received his Ph.D. in Botany from the University of California at Davis in 1982. Currently he is Associate Professor and past Chair of Biology at Mills College. His research emphasizes the ecology and physiology of plants native to western North America, and he has a special interest in the conservation of endangered species. Dr. Pavlik is senior author of *Oaks of California*—an elegant book on the ecology, diversity, and value of oaks written for a general audience—and he has written many technical journal articles which have an international readership of professional biologists. Bruce lives in Oakland with his wife, Lynn, and two sons, Ben and Matt.

Dr. Susan Lindstrom first developed her interest in native plant uses while working as a naturalist in the north-central Sierra Nevada, subsequent to receiving her bachelor's degree in anthropology from the University of California at Berkeley in 1972. She furthered hits research pursuit throughout her graduate studies at the University of California at Davis, completing a Ph.D. in Anthropology in 1992. Lindstrom continues to live and work in the Lake Tahoe area, consulting widely with federal and state governments and private industry on archaeologic, historic and heritage preservation concerns.

Frank Drysdale lives in western Oregon where he conducts independent research, gardens, and works on regional environmental issues. He is a graduate of California Polytechnic University at Pomona (BS), Utah State University (MS), and the University of California at Davis (Ph.D. in Botany, 1971). He is coauthor of *Coastal Ecology: Bodega Head* (University of California Press), and *The Energetics of the United States of America: An Atlas* (US Department of Energy/ NTIS).

CREDITS

MAPS AND ILLUSTRATIONS

Kat Anderson: 163 (both)
Jean Brennan-Hagen: 33, 80, 167
Michal Yuval: 4, 14, 26, 39, 56, 72, 74, 90, 100, 103, 132, 133, 140
Margaret Warriner Buck in Parsons, *The Wild Flowers of California*, 1930: 168, 172, 175, 178, 179
Paul Landacre in Peattie, *A Natural History of Western Trees*: 65, 112
Jepson Manual: 26 upper left
Alfred Marty: 200
Erwin Raisz: *viii*
Leslie Randall: 160
Sarah Young: 22, 53, 70, 99, 130

PHOTOGRAPHS

Walt Anderson: x, 21, 23, 27, 27, 29, 30, 31, 46, 53, 60, 72, 73, 75, 77, 79, 82 middle and bottom left, 85, 122, 142, 143, 157, 161, 173 top and middle
Carol Arnold: 198
Michael Barbour: 5 bottom, 15 (both), 62, 66, 94, 108, 113, 114, 125, 177, 200 (both)
David Cavagnaro: Title page, 5 top, 19, 20, 59, 71, 82 right and top left, 84 (both), 87, 89, 96, 134, 156, 171 top and middle, 185 (both), 189, 195, 200 top, 204
N.H. (Dan) Cheatham: 117
Susan Cochrane: 132, back cover inset
Phyllis Faber: 40, 45, 49, 116 bottom
William Follette: 28, 64, 65, 101, 121, 127, 173 bottom, 181
Pliney E. Goddard, courtesy of the Lowie Museum, University of California, Berkeley: 171 bottom
Mary Ann Griggs: 184, 196 (both)
D. Guiliani: 158
Robert Haller: Cover, 2, 8, 10, 52, 81, 123, 144, 152, 193
J. Harris: 37
John Hudson, courtesy of the Field Museum of Natural History, Chicago, IL: 166
Verna Johnston: 25, 86, 102 left
Steve Junak: 197
Todd Keeler-Wolf: 78
Courtesy of the Lowie Museum, University of California, Berkeley: 174
C. Hart Merriam, courtesy of the Bancroft Library, University of California, Berkeley: 162
Bruce Pavlik: 12, 35, 38, 42, 43, 63, 98, 102 right, 104, 116 top, 119, 131, 135, 136, 138, 141, 143, 146, 148, 154
Oren Pollak: 118, 138
San Diego Historical Society, Ticor Collection: 170
Lynn Suer: 245 upper middle